AQA GCSE (9–1) Physics topics for Combined Science

Foundation Support Workbook

Penny Johnson

Beverly Rickwood

William Collins' dream of knowledge for all began with the publication of his first book in 1819. A self-educated mill worker, he not only enriched millions of lives, but also founded a flourishing publishing house. Today, staying true to this spirit, Collins books are packed with inspiration, innovation and practical expertise. They place you at the centre of a world of possibility and give you exactly what you need to explore it.

Collins. Freedom to teach

HarperCollins Publishers
The News Building
1 London Bridge Street
London SE1 9GF

Browse the complete Collins catalogue at
www.collins.co.uk

First edition 2016

10 9 8 7 6 5 4 3 2 1

© HarperCollins Publishers 2016

ISBN 978-0-00-818956-3

Collins® is a registered trademark of HarperCollins Publishers Limited

www.collins.co.uk

A catalogue record for this book is available from the British Library

Commissioned by Gillian Lindsey
Project managed by Sarah Thomas
Copy edited by Jane Roth
Proofread by Heather Addison
Technical review by Christine Graham
Typeset by Jouve India Pvt Ltd.,
Cover design by We are Laura and Jouve
Cover image: 123rf/reddz
Printed by Martins the Printers, Berwick upon Tweed

The publisher would also like to thank Linda Needham and Richard Needham for their support in the development of this book.

Introduction

· ·

This workbook will help you build your confidence in answering Physics questions for GCSE Combined Science, Foundation tier.

It gives you practice in using key scientific words, writing longer answers and applying maths and practical skills.

The opening summary shows what definitions and equations you need to learn.

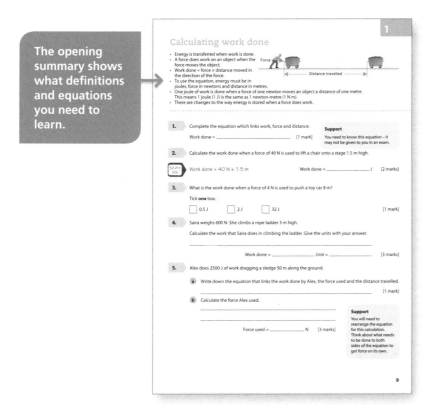

Learn how to answer test questions by seeing part of the answer filled in.

This will help you develop the skills you need to write longer answers, or to use maths in science.

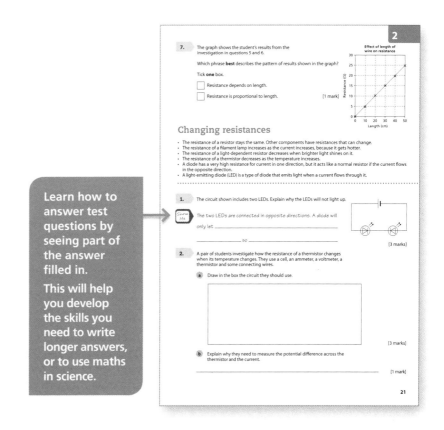

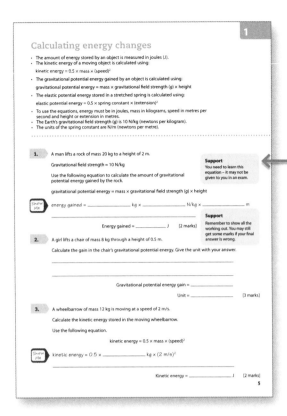

The amount of support gradually decreases throughout the workbook. As you build your skills you should be able to complete more of the questions yourself.

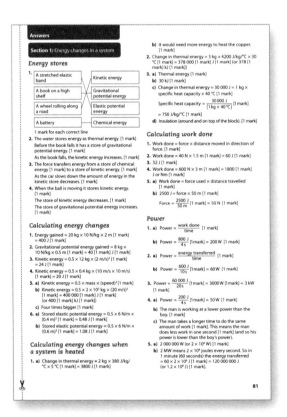

There are answers to all the questions at the back of the book. You can check your answers yourself or your teacher might tear them out and give them to you later to mark your work.

Contents

Energy stores

- Energy can be stored and also transferred from one store to another.
- Energy stores include elastic (in a stretched spring), gravitational (when a weight is lifted), kinetic (in a moving object), chemical (in a fuel) and thermal (energy stored in a warm object).
- Energy can be transferred from one energy store to another. For example, energy is transferred by heating, by electric current in a circuit, and when a force moves through a distance.

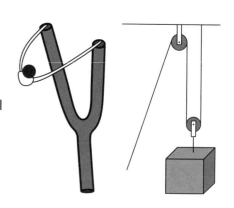

---

1. Draw **one** line from each object to the correct energy store.

Show Me

A stretched elastic band	Kinetic energy
A book on a high shelf	Gravitational potential energy
A wheel rolling along a road	Elastic potential energy
A battery	Chemical energy

[4 marks]

2. Complete the sentences below.

Choose words from the box. Use a word only once. You will not need to use all the words.

| chemical   thermal   kinetic   gravitational potential   elastic potential |

When you use a kettle to boil water, energy is transferred to the water. The water stores energy as

_____ energy.

When a book falls off a shelf it accelerates. Before the book falls it has a store of _____

energy. As the book falls, the _____ energy increases.   [3 marks]

3. The sentences below describe how different stores of energy change as a car accelerates or slows down.

Complete the sentences.

The car's engine burns a fuel. The engine applies a force to the car, which accelerates.

The force transfers energy from a store of _____ to a store of

_____.

As the car slows down the amount of energy in the kinetic store _____.

[3 marks]

4. When a ball is thrown up in the air, energy is transferred from one store to another.

Describe how the energy is transferred between the stores as the ball travels up and slows down.

_____

_____

[3 marks]

Support

There are 3 marks here so you need to make three points. You must say more than just what the original and final energy stores are.

4

Calculating energy changes

- The amount of energy stored by an object is measured in joules (J).
- The kinetic energy of a moving object is calculated using:

  kinetic energy = 0.5 × mass × (speed)^2

- The gravitational potential energy gained by an object is calculated using:

  gravitational potential energy = mass × gravitational field strength (g) × height

- The elastic potential energy stored in a stretched spring is calculated using:

  elastic potential energy = 0.5 × spring constant × (extension)^2

- To use the equations, energy must be in joules, mass in kilograms, speed in metres per second and height or extension in metres.
- The Earth's gravitational field strength (g) is 10 N/kg (newtons per kilogram).
- The units of the spring constant are N/m (newtons per metre).

1. A man lifts a rock of mass 20 kg to a height of 2 m.

   Gravitational field strength = 10 N/kg

   Use the following equation to calculate the amount of gravitational potential energy gained by the rock.

   gravitational potential energy = mass × gravitational field strength (g) × height

   Support
   You need to learn this equation – it may not be given to you in an exam.

   Show Me   *energy gained =* _____ kg × _____ N/kg × _____ m

   _____

   Energy gained = _____ J   [2 marks]

   Support
   Remember to show all the working out. You may still get some marks if your final answer is wrong.

2. A girl lifts a chair of mass 8 kg through a height of 0.5 m.

   Calculate the gain in the chair's gravitational potential energy. Give the unit with your answer.

   _____

   _____

   Gravitational potential energy gain = _____

   Unit = _____   [3 marks]

3. A wheelbarrow of mass 12 kg is moving at a speed of 2 m/s.

   Calculate the kinetic energy stored in the moving wheelbarrow.

   Use the following equation.

   kinetic energy = 0.5 × mass × (speed)^2

   Show Me   *kinetic energy = 0.5 ×* _____ kg × (2 m/s)^2

   _____

   Kinetic energy = _____ J   [2 marks]

4. A student wants to calculate the kinetic energy of a football. The ball has mass 0.4 kg and travels at a speed of 10 m/s. This is his answer:

$$\text{kinetic energy} = 0.5 \times \text{mass} \times (\text{speed})^2$$

$$= 0.5 \times 0.4 \times 10 = 2\,J$$

The student has made a mistake. Write down the correct calculation and answer.

_____

_____

Kinetic energy = _____ J   [2 marks]

5. **a** Write down the equation that links kinetic energy, mass and speed.

Equation: _____   [1 mark]

b A van, mass 2×10^3 kg, is travelling at a speed of 20 m/s.

Calculate the kinetic energy of the van. Give the unit with your answer.

_____

_____

Kinetic energy = _____ Unit = _____   [3 marks]

c The van's speed doubles. How many times bigger is the car's kinetic energy?

Tick **one** box.

☐ the same   ☐ twice as big   ☐ four times bigger   [1 mark]

6. A spring has a spring constant of 6 N/m.

a Calculate how much elastic potential energy the spring stores when it is extended by 0.4 m.

Use the following equation.

$$\text{elastic potential energy} = 0.5 \times \text{spring constant} \times (\text{extension})^2$$

Show Me

$$\text{stored elastic potential energy} = 0.5 \times 6\,\text{N/m} \times (0.4\,\text{m})^2$$

$$= 0.5 \times 6 \times \text{_____}$$

Stored energy = _____ J   [2 marks]

> **Support**
> Always check that you put values into the equation using SI units. Distance must be in metres. Change the values to SI units before starting any of the calculations.

b Calculate how much elastic potential energy the spring stores if it is extended by 60 cm.

_____

_____

Stored energy = _____ J   [2 marks]

Calculating energy changes when a system is heated

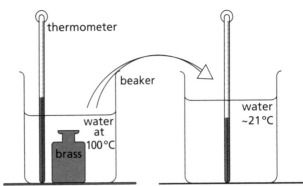

- A system is an object or group of objects.
- If the temperature of the system increases, the increase in temperature depends on the type of substance, the mass of the substance heated and the energy input to the system.
- The amount of energy stored in or released from a system as its temperature changes can be calculated using:

  change in thermal energy = mass × specific heat capacity × temperature change

- The specific heat capacity of a substance is the amount of energy needed to raise the temperature of one kilogram of the substance by one degree Celsius.
- To use the equation, energy must be in joules, mass in kilograms and temperature in degrees Celsius, °C.

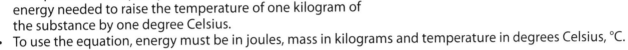

1. Copper has a specific heat capacity of 380 J/kg/°C.

a Calculate the amount of energy needed to raise the temperature of a 2 kg block of copper by 5 °C.

Use the following equation.

change in thermal energy = mass × specific heat capacity × temperature change

Change in thermal energy = _____ kg × _____ J/kg/°C × _____ °C

Change in thermal energy = _____ J   [2 marks]

b Aluminium has a specific heat capacity of 900 J/kg/°C.

Compare the amount of energy needed to raise the temperature of a 2 kg block of aluminium by 5 °C with your answer to (a).

Tick **one** box.

☐ It would need more energy to heat the aluminium.

☐ It would need more energy to heat the copper.

☐ It would take the same amount of energy to heat both.   [1 mark]

2. Water has a specific heat capacity of 4200 J/kg/°C.

Calculate the amount of energy that has to be supplied to 3 kg of water to raise its temperature from 20 °C to 50 °C. Include the correct unit in your answer.

_____

_____

Change in thermal energy = _____

Unit = _____

> **Support**
> Always show your working out. You may still get some marks if your final answer is wrong.

[3 marks]

3. A student is doing an experiment to measure the specific heat capacity of metal **X**. She uses the equipment shown here.

This is the method used.

1 Use the thermometer to measure the temperature of the metal block.

2 Switch on the heater and heat the metal block for 10 minutes.

3 Switch the heater off and take the temperature again.

4 Calculate how much the temperature of the block has risen.

thermometer

50 W Heater

block of metal X

a In this experiment the electric current transfers energy to the metal block.

State the type of energy store in the block.

_____

[1 mark]

b The heater supplies 30 000 J of energy.

Convert 30 000 J to kilojoules.   30 000 J = _____ kJ   [1 mark]

c The block has a mass of 1 kg. The temperature of the block rose from 20 °C to 60 °C. The heater supplied 30 000 J of energy.

Calculate the specific heat capacity of metal **X**.
Use the following equation.

change in thermal energy = mass × specific heat capacity × temperature change

_____

_____

_____

Specific heat capacity = _____ J/kg/°C   [3 marks]

> **Support**
>
> You will need to rearrange the equation for this calculation. To find specific heat capacity you will need to rearrange the equation so specific heat capacity is on its own, on one side of the equals sign.
>
> The equation is given in the form $a = b \times c \times d$. To find c, divide both sides of the equation by b. Then divide both sides of the equation by d.

d Some of the energy transferred by the electric current does not raise the temperature of the water. This causes an error in the result.

Suggest **one** improvement to the experiment to reduce this error and make the calculated value more accurate.

_____

_____

_____   [1 mark]

Calculating work done

- Energy is transferred when work is done.
- A force does work on an object when the force moves the object.
- Work done = force × distance moved in the direction of the force.
- To use the equation, energy must be in joules, force in newtons and distance in metres.
- One joule of work is done when a force of one newton moves an object a distance of one metre. This means 1 joule (1 J) is the same as 1 newton-metre (1 N m).
- There are changes to the way energy is stored when a force does work.

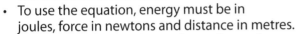

Force

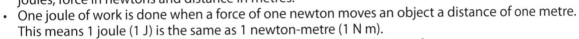

Distance travelled

1. Complete the equation which links work, force and distance.

   Work done = _____. [1 mark]

   Support

   You need to know this equation – it may not be given to you in an exam.

2. Calculate the work done when a force of 40 N is used to lift a chair onto a stage 1.5 m high.

   Show Me

   Work done = 40 N × 1.5 m   Work done = _____ J   [2 marks]

3. What is the work done when a force of 4 N is used to push a toy car 8 m?

   Tick **one** box.

   ☐ 0.5 J   ☐ 2 J   ☐ 32 J   [1 mark]

4. Saira weighs 600 N. She climbs a rope ladder 3 m high.

   Calculate the work that Saira does in climbing the ladder. Give the units with your answer.

   _____

   Work done = _____ Unit = _____   [3 marks]

5. Alex does 2500 J of work dragging a sledge 50 m along the ground.

   a. Write down the equation that links the work done by Alex, the force used and the distance travelled.

   _____ [1 mark]

   b. Calculate the force Alex used.

   _____

   _____

   Force used = _____ N   [3 marks]

   Support

   You will need to rearrange the equation for this calculation. Think about what needs to be done to both sides of the equation to get force on its own.

Power

- Power means the rate at which energy is transferred or the rate at which work is done.
- Power = $\frac{\text{work done}}{\text{time}}$ or power = $\frac{\text{energy transferred}}{\text{time}}$
- Power is measured in watts (W) or kilowatts (kW). 1 kW = 1000 W.
- An energy transfer of 1 joule per second is equal to a power of 1 watt.

1. **a** Write down the equation which links work and power.

Power = _____

[1 mark]

b A man does 800 J of work in 4 seconds. Calculate his power.

_____

Power = _____ W   [2 marks]

2. The electric current in a light bulb transfers energy to give light. Some thermal energy is also released.

a Write down the equation which links energy transferred, power and time.

_____

[1 mark]

b The light bulb transfers 600 J of energy in 10 seconds. Calculate the power of the light bulb.

Power = $\frac{600\,J}{10\,s}$

Power = _____ W   [2 marks]

3. Calculate the power of an electric kettle that transfers 60 000 J of electrical energy in 20 seconds. Give your answer in kW.

_____

Support
k means kilo which is 1000 or 10^3. 1 kW is 1000 W.

_____

Power = _____ kW   [3 marks]

4. **a** A boy does 200 J of work in 4 seconds. Calculate his power.

_____

Power = _____ W   [2 marks]

b A man takes 10 seconds to do the same amount of work as the boy in part (a). Which **one** of the statements below is correct?

Tick **one** box.

☐ The man is working at a lower power than the boy.

☐ They are both working at the same power.

☐ The man is working at a higher power than the boy. [1 mark]

c Give a reason for your answer to part (b).

Show Me

The man takes a longer time to do the same amount of work.
This means the man does less work in one second and so his power is lower than the boy's power.

_____

_____ [2 marks]

5. A large wind turbine can generate electrical power at 2 MW.

a State the power in watts that 2 MW equals.

Power = _____ W   [1 mark]

> **Support**
> M means mega which is 1 000 000 or 10^6. 1 MW is 1 000 000 W.

b Calculate the energy that this turbine transfers in one minute. Give your answer in joules (J).

Show Me

2 MW means 2 million joules every second. So in 1 minute (60 seconds) the energy transferred = 60 × 2 × 1 000 000 J

Energy transferred in one minute = _____ J   [2 marks]

> **Support**
> You could also show your working and answer in standard form. In 60 s the energy transferred = $60 \times 2 \times 10^6$ J. In your final answer, remember that in standard form the number multiplied by the power of ten must be between 1 and 10. For example, 80×10^6 is 8×10^7 in standard form.

Conservation of energy

- Energy cannot be created or destroyed.
- Energy can be transferred usefully, stored or dissipated. The total amount of energy does not change.
- Dissipated energy has been transferred to a store where it cannot be used. For example, when a hot toaster heats the surrounding air. The energy stored in the air has increased, but this energy is wasted.

60 J

0.2 J useful energy to give light

59.8 J energy heats the air

1. Which one of these statements is **not** true?

Tick **one** box.

☐ Energy can be destroyed.

☐ Energy can be dissipated.

☐ Energy can be stored.

[1 mark]

2. Complete the sentences below.

Choose words from the box. Use a word only once. You will not need all the words.

chemical	kinetic	potential	thermal

A vacuum cleaner is designed to transfer energy from the mains electrical supply to

_____ energy.

When a vacuum cleaner is used some of the input energy is wasted by transfer of

_____ energy to the surroundings.

[2 marks]

3. The electric current in a fluorescent light transfers energy. For every 100 J of input energy, only 60 J is released as light.

Explain what happens to the rest of the input energy.

_____

_____

_____ [3 marks]

Support

There are 3 marks here so you need to make three points.

Remember that 'explain' means you have to say **what** happens and also **why** it happens.

4. The diagram shows an example of energy transfer in a wind turbine.

a Give the amount of output energy that is transferred by an electric current.

Energy transferred by current =

_____ J [1 mark]

2000 J Kinetic energy

Energy transferred to surroundings 800 J

Energy transferred by an electric current

b Give **one** way that the wind turbine is likely to waste energy.

_____ [1 mark]

12

Ways of reducing unwanted energy transfers

- Some energy transfers are wasteful. We can try to reduce the waste.
- Unwanted thermal energy transfer from moving parts can be reduced by lubrication to reduce friction.
- Unwanted thermal energy transfer from hot objects can be reduced by using thermal insulation.
- The higher the thermal conductivity of a material, the higher the rate of energy transfer by conduction through the material.
- The rate of cooling of a building is affected by the thickness and the thermal conductivity of its walls.

roof 25%
windows 10%
walls 35%
doors 15%
floors 15%

---

1. Which **one** of these statements is correct? Tick **one** box.

☐ We can stop energy transfers happening.

☐ Lubrication reduces unwanted energy transfers.

☐ Using thermal insulation increases energy transfer.

[1 mark]

2. The table lists some changes that could be made to a house.

Complete the table to show what effect each change would have on the thermal energy transfer out of the house.

Put **one** tick in each row.

Suggested change	Increase thermal transfer	Reduce thermal transfer
Make the outside walls thicker	☐	☐
Remove loft insulation	☐	☐
Replace carpets with stone floors	☐	☐
Add double glazing to the windows	☐	☐

[4 marks]

3. Complete the sentences below.

Use the words from the box. Use each word once.

friction	kinetic	lubricated	thermal

Electric motors are designed to transfer energy from the electrical supply to _____ energy.

Not all the input energy is usefully transferred. There is _____ between moving parts in the motor.

This means that the motor transfers some energy to the surroundings as _____ energy.

The motor can be _____ to reduce the amount of unwanted energy transfer.

[4 marks]

13

Efficiency

- The efficiency of a device tells us how much of the energy going into it is transferred to a useful output.
- The higher the efficiency, the less energy is dissipated (wasted).
- The energy efficiency for any energy transfer can be calculated using the equation:

$$\text{efficiency} = \frac{\text{useful output energy transfer}}{\text{useful input energy transfer}}$$

- This equation can also be used for efficiency:

$$\text{efficiency} = \frac{\text{useful power output}}{\text{total power input}}$$

- Efficiency can be written as a decimal or a percentage.
- It is not possible for a system to have an efficiency greater than 1 (or 100%).

1. The efficiency of a lamp is 40%.

> **Support**
> To convert from a percentage to a decimal, divide the percentage by 100.

a Write 40% as a decimal.

 $40\% = \dfrac{40}{100}$

40% = _____ [1 mark]

b Write 0.65 as a percentage.

_____

0.65 = _____% [1 mark]

2. The electric current in a fluorescent light transfers energy. For every 100 J of input energy, 60 J is output as light.

Calculate the efficiency of this energy transfer. Give your answer as a percentage.

> **Support**
> To convert from a fraction to a percentage, first change the fraction to a decimal. Then multiply by 100.

 $\text{Efficiency} = \dfrac{60\,\text{J}}{100\,\text{J}}$

Efficiency = _____% [2 marks]

3. An energy transfer diagram for a washing machine is shown.

Calculate the efficiency of this washing machine. Give your answer as a decimal.

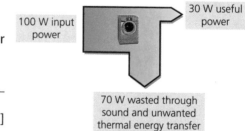

_____

Efficiency = _____ [2 marks]

4. A man is moving some furniture. He does 8000 J of work. Only 6000 J of the work done is useful in moving the furniture.

Calculate the efficiency of this energy transfer. Give your answer as a percentage.

_____

Efficiency = _____% [2 marks]

Renewable and non-renewable energy resources

- Fuels such as coal, oil, gas and nuclear fuel are not renewable. Supplies will run out.
- Renewable energy resources can be replaced as they are used.
- Renewable energy resources include bio-fuel, wind, hydroelectricity, geothermal, the tides, the Sun and water waves.
- Some energy resources can be used to generate electricity.
- Most renewable resources do not generate a predictable (reliable) amount of electrical power.
- There is a different type of impact on the environment from using different energy resources.

1. Complete the table to show whether each type of energy resource is renewable or non-renewable.

Tick **one** box in each row.

Resource	Renewable	Non-renewable
Bio-fuel	☐	☐
Tides	☐	☐
Coal	☐	☐
Wind	☐	☐

[4 marks]

2. A student wrote the statement below about renewable fuels in a test.

Renewable energy resources can be recycled.

The answer was marked as incorrect.

Write a **correct** definition of renewable energy resources.

_____ [1 mark]

3. Draw **one** line from each energy resource to the environmental impact it may have.

bio-fuel	Increases carbon dioxide levels in the atmosphere
coal	Creates visual and noise pollution
tides	Requires removal of forests and areas for growing food crops
wind	Requires barriers which destroy wildlife habitats

[4 marks]

4. Explain why some renewable resources are not available all the time.

_____

_____ [2 marks]

5. This chart shows how the use of energy resources in one country has changed over the last 25 years.

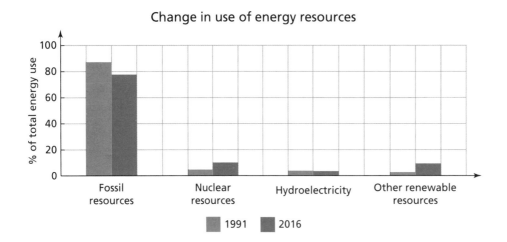

Describe what the chart tells you about the following:

a The use of fossil fuels

_____ [1 mark]

b The use of nuclear fuels

_____ [1 mark]

c The overall use of renewable energy

_____ [1 mark]

6. A car-hire company has 200 cars. At present, 150 of its cars run on petrol or diesel.
The rest run on bio-fuel.

The company wants to replace its petrol and diesel cars with cars that run on bio-fuels.

a What fraction of its cars run on bio-fuel?

_____ [1 mark]

b What is the ratio of cars that run on bio-fuel to cars that run on petrol or diesel?

Tick **one** box.

☐ 1:4   ☐ 1:3   ☐ 3:1   [1 mark]

c Give **one** advantage of using bio-fuel rather than petrol or diesel to run cars.

_____ [1 mark]

d Describe **one** other use of bio-fuels.

_____ [1 mark]

Circuit diagrams

- We use standard symbols to represent different components in electric circuits.
- Circuit diagrams show how the components are linked together. The wires are always drawn as straight lines.

1. Describe what these two symbols show.

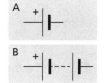

 A: Shows a cell.

B: Shows a _____, which is two or

more _____.

[2 marks]

2. Draw the symbol for each of these components in the boxes.

ammeter	voltmeter	resistor	bulb	variable resistor

[5 marks]

3. Draw a line from each symbol to the name of the component.

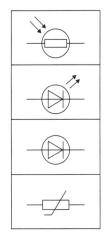

thermistor
light-dependent resistor (LDR)
diode
light-emitting diode (LED)

Support
Remember that this
⎓ represents a
resistor, so any symbol that
includes this is a kind of
resistor. A thermistor is a
type of resistor. Two arrows
on a symbol represent light.

[4 marks]

4. A circuit is needed to measure the current through a light-emitting diode.

Draw the circuit diagram in the box.

[3 marks]

Electrical charge and current

- Current flows in a complete circuit only if there is a potential difference (voltage).
- The current is the same everywhere in a series circuit.
- Electric current is a flow of electric charge. Charge is measured in coulombs (C).
- Charge flow, current and time are linked by this equation:

charge flow = current × time

- To use the equation, charge must be in coulombs, current in amps and time in seconds

1. Name a circuit component used to provide a potential difference in a circuit.

_____ [1 mark]

2. The diagram shows a series circuit. Ammeter A1 reads 2 A.

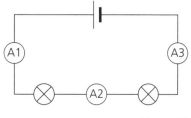

a State what the readings will be on ammeters A2 and A3.

A2: _____

A3: _____ [1 mark]

b Explain your answers to part (a).

_____ [1 mark]

3. Current, charge and time are linked by this equation:

charge flow = current × time

> **Support**
> You need to learn this equation – it may not be given to you in an exam.

Write down the units for each quantity in this equation. Write the word and its abbreviation.

charge flow: _____

 current: *amp, A*

time: _____ [3 marks]

4. A circuit has a current of 3 amps.

a Calculate the charge that flows in the circuit in 20 seconds.

 Charge flow = current × time

= 3 A × 20 s

Charge flow = _____ C [2 marks]

b The current of 3 A flows in the circuit for 10 minutes.

Calculate the charge that flows.

10 minutes × 60 = 600 seconds

Charge flow = _____ A × _____ s

Charge flow = _____ C

Support

The SI unit for time is the second. Convert minutes to seconds by multiplying by 60.

[3 marks]

5. The current in a circuit is 5 A. Calculate how long it takes for 25 000 C to flow around the circuit.

25 000 C = _____ A × time

$$\text{Time} = \frac{25\ 000\ C}{_____}$$

Time = _____ s   [3 marks]

6. Calculate the charge that flows when there is a 5 mA current in a circuit for 20 minutes.

$$5\ mA = \frac{5}{1000} = 0.005\ A$$

20 minutes = 20 × 60 s = _____ s

_____

Charge flow = _____ C   [4 marks]

Support

The SI unit for current is the amp. There are 1000 milliamps (mA) in 1 amp. Convert mA to A by dividing by 1000.

Electrical resistance

- Resistance is a way of saying how difficult it is for electricity to flow through a component.
- Resistance is measured in ohms (Ω).
- The size of a current depends on the potential difference and the resistance.
- Current, potential difference and resistance are linked by the equation:

potential difference = current × resistance

- This equation can also be written as:

$$V = I \times R$$

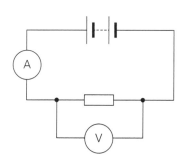

1. Complete the sentences below using words from the box.

coulombs	current	voltage	increases	decreases

The potential difference is sometimes called the _____.

If the potential difference in a circuit is increased, the current _____.

If the resistance is increased for the same potential difference, the current _____.

[3 marks]

2. A circuit has a resistance of 20 Ω.

Calculate the potential difference needed to give a current of 4 A in the circuit.

Show Me

Potential difference = current × resistance

= _____ A × _____ Ω

Potential difference = _____ V   [3 marks]

3. A cell in a circuit provides a potential difference of 12 V. The current in the circuit is 3 A.

Calculate the resistance of the circuit. Give the units with your answer.

Show Me

12 V = 3 A × resistance

Resistance = $\dfrac{12\ V}{3\ A}$

Resistance = _____ Unit _____   [4 marks]

4. A student sets up this circuit to find out how the resistance of a wire changes with the length of the wire.

The student has not set up the circuit correctly.

Explain what needs changing.

_____

_____

_____ [3 marks]

5. The student from question 4 now uses the corrected circuit to investigate the relationship between the length of the wire and its resistance.

a State the independent variable in this investigation. _____   [1 mark]

b State the dependent variable in this investigation. _____   [1 mark]

c Name **one** control variable in this investigation. _____   [1 mark]

6. Describe the method that the student in questions 4 and 5 should use to investigate how the length of a wire affects its resistance.

Your answer should state how enough data will be collected to show the relationship.

_____

_____

_____ [3 marks]

7. The graph shows the student's results from the investigation in questions 5 and 6.

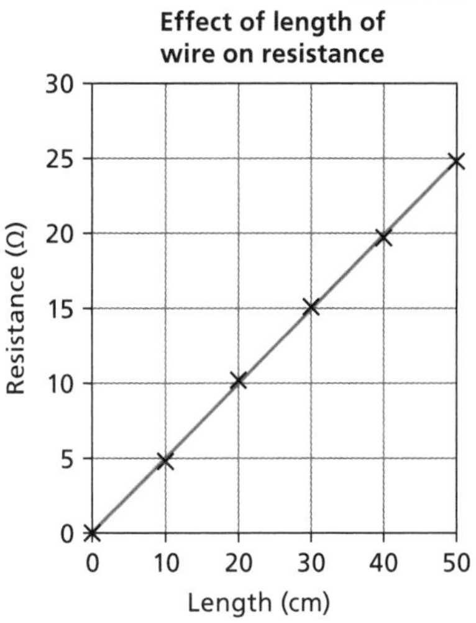

Which phrase **best** describes the pattern of results shown in the graph?

Tick **one** box.

☐ Resistance depends on length.

☐ Resistance is proportional to length. [1 mark]

Effect of length of wire on resistance

Changing resistances

- The resistance of a resistor stays the same. Other components have resistances that can change.
- The resistance of a filament lamp increases as the current increases, because it gets hotter.
- The resistance of a light-dependent resistor decreases when brighter light shines on it.
- The resistance of a thermistor decreases as the temperature increases.
- A diode has a very high resistance for current in one direction, but it acts like a normal resistor if the current flows in the opposite direction.
- A light-emitting diode (LED) is a type of diode that emits light when a current flows through it.

1. The circuit shown includes two LEDs. Explain why the LEDs will not light up.

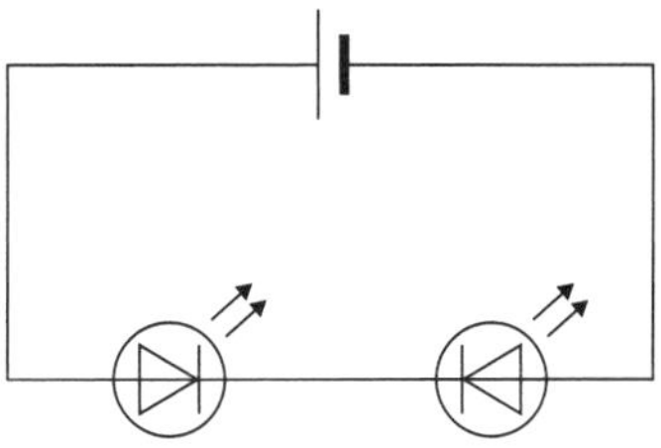

Show Me

The two LEDs are connected in opposite directions. A diode will

only let _____

_____ so _____

[3 marks]

2. A pair of students investigate how the resistance of a thermistor changes when its temperature changes. They use a cell, an ammeter, a voltmeter, a thermistor and some connecting wires.

a Draw in the box the circuit they should use.

[3 marks]

b Explain why they need to measure the potential difference across the thermistor and the current.

_____ [1 mark]

3. Complete the sentences using words from the box. You may need to use a word more than once.

brightness	decreases	increases	temperature

a When the potential difference across a filament lamp _____,

its resistance _____. [1 mark]

b When the _____ of a thermistor increases,

its resistance _____. [1 mark]

c When the _____ decreases, the resistance of

an LDR _____. [1 mark]

4. A simple fire alarm sounds a buzzer if the temperature gets too high.
These sentences explain how it works.

Number the sentences to put them in a good order.

[ ] If there is a fire the temperature rises and so the resistance of the thermistor goes down.

[ ] When the thermistor is cold its resistance is high.

Show Me ▶ [1] The circuit includes a cell, a thermistor and a buzzer.

[ ] This allows a bigger current to flow in the circuit, so the buzzer sounds.

[ ] Only a very small current can flow in the circuit, so the buzzer does not sound. [5 marks]

5. A fridge has a thermostat that turns the cooling system on if the temperature in the fridge gets too high.

A thermistor can be used in a circuit to keep the temperature in the fridge constant.

Explain how the circuit works.

Show Me ▶ When the temperature is low, the resistance _____

_____

If the temperature rises, _____

_____ [4 marks]

6. A thermistor is used in a heater circuit to keep a house warm.

Explain why the heater needs a special switch that turns the heater **on** when the current is **low**.

> **Show Me**
>
> A thermistor has a low resistance when the temperature is _____

_____

_____ [2 marks]

7. Most street lights have a sensor that switches the lights on when it gets dark.

Describe how an LDR can be used in a sensor circuit to control street lights.

_____

_____

_____ [3 marks]

Series and parallel circuits

- In a series circuit all the components are on a single loop. In a parallel circuit there is more than one path that the current can take.
- In a series circuit the current is the same through all the components. The potential difference of the power supply is shared between the components.
- In a parallel circuit the potential difference across each branch of the circuit is the same. The current through the cell is the sum of the currents through the separate branches.
- If two components are used in series, the total resistance is the sum of the resistance of each component.

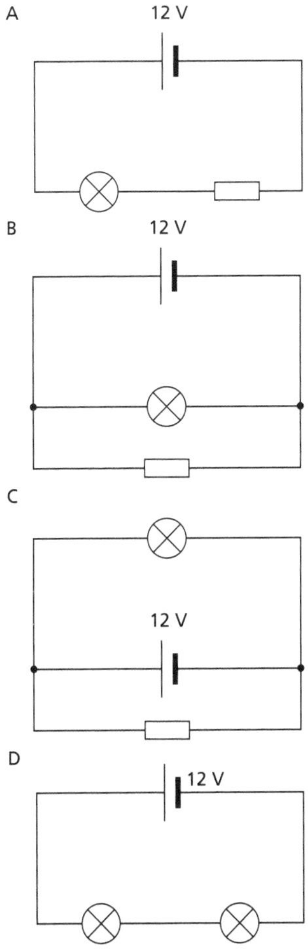

1. Look at the circuits in the diagram.

Write down the letters for the circuits that are:

a series circuits: _____ [1 mark]

b parallel circuits: _____ [1 mark]

2. In circuit A the potential difference across the resistor is 8 V.

a State the potential difference across the bulb.

Potential difference = _____ V [1 mark]

b Explain your answer to part (a).

> **Show Me**
>
> In a series circuit the potential difference is

_____. [1 mark]

A 12 V

B 12 V

C 12 V

D 12 V

3. In circuit D the two bulbs are identical.

State the potential difference across each bulb.

Potential difference = _____ V   [1 mark]

4. In circuit A the current through the resistor is 1.5 A.

a Give the value of the current through the bulb.

Current = _____ A   [1 mark]

b Explain your answer to part (a).

Show Me

In a _____ circuit the current

is _____.   [1 mark]

Support

Identical bulbs in series will have the same potential difference across each one.

5. **a** Give the value of the potential difference across the resistor in circuit B.

Potential difference = _____ V   [1 mark]

b Explain your answer to part (a).

Show Me

In a parallel circuit the potential difference across each branch is _____

[1 mark]

6. In circuit B the current through the cell is 3 A, and the current through the bulb is 1 A.

a State the current through the resistor.

Current = _____ A   [1 mark]

b Explain your answer to part (a).

_____

_____   [1 mark]

7. In circuit D each bulb has a resistance of 6 Ω. What is the total resistance in the circuit?

Resistance = _____ Ω   [1 mark]

Mains electricity

- In the UK the mains electricity supply is alternating current with a frequency of 50 Hz and a potential difference of about 230 V.
- The live (brown) wire carries the alternating potential difference, the neutral (blue) wire completes the circuit, and the earth (green and yellow) wire is for safety.
- The live wire is at a potential difference of 230 V compared to earth (0 V). The neutral and earth wires are normally at 0 V. The earth wire only carries a current if there is a fault.
- A live wire may be dangerous at any time. A connection between a live wire and earth will cause a large current to flow which could start a fire or give an electric shock.

1. **a** Describe the difference between an **alternating** potential difference and a **direct** potential difference.

Show Me

A direct potential difference always acts in the _____.

With an alternating potential difference the direction _____

_____. [2 marks]

b Name a circuit component that supplies a direct potential difference.

_____ [1 mark]

2. The diagram shows the inside of a three-pin plug.
Each pin is connected to one of the wires inside a cable.

Complete the following with the name of each wire.

The wire connected to:

A is the _____ wire.

B is the _____ wire.

C is the _____ wire.
[3 marks]

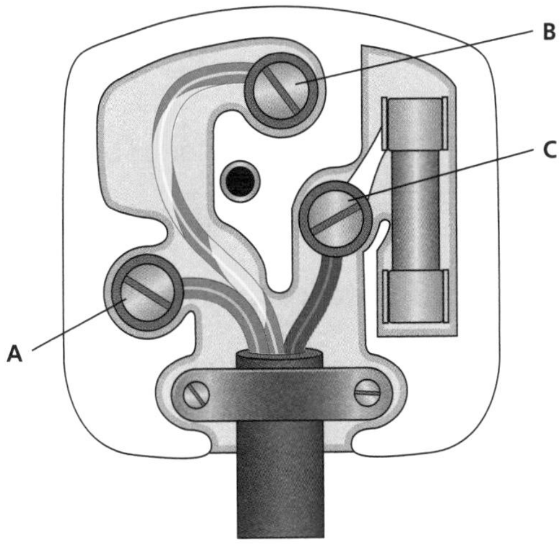

3. A fault in an appliance may lead to the live wire touching a metal casing.

a Describe what will happen in the earth wire if this happens.

_____ [1 mark]

b Explain what would happen to a person touching the live casing, if there were no earth wire.

Show Me

The person would provide a connection between the live

supply and _____.

The live wire is at a potential difference of _____

compared to earth, so _____

_____ [4 marks]

Support

Remember that if there are 4 marks for a written question, you need to make four different points in your answer.

Energy changes in circuits

- Domestic (household) appliances transfer energy from cells or from the mains supply to kinetic energy or by heating.
- The power rating of an appliance tells us how many joules of energy the appliance transfers each second.
- The work done (amount of energy transferred) by an appliance depends on the power of the appliance and how long it is switched on.
- The energy transferred is linked to the power and the time by this equation:

energy transferred = power × time

- The energy transferred can also be calculated from the charge flow and the potential difference:

energy transferred = charge flow × potential difference

1. A kettle with power rating 2 kW transfers energy to the water inside it.

   State the energy transferred every second. Give your answer in joules.

   Energy transferred every second = _____ J   [1 mark]

   Support

   Remember that 1 kW = 1000 W, and that 1 W means 1 J transferred each second.

2. Complete the sentences below using words from the box. You will not need to use all the words.

energy	greater	higher potential difference
lower potential difference	power	a longer time a shorter time

   The energy transferred by an appliance is greater if the appliance has a higher _____

   or if it is switched on for _____ .

   The work done (energy transferred) by an electric current is _____

   if more charge flows or if the charge is 'pushed' by a _____ . [4 marks]

3. State the standard unit for each of these quantities.

   Show Me

   charge: *coulomb, C*

   power: _____

   energy: _____

   time: _____   [4 marks]

4. Calculate the energy transferred when 1 coulomb of charge flows through a bulb with a potential difference of 230 V across it.

   Support
   You need to know this equation – it may not be given to you in an exam.

   Show Me

   Energy transferred = charge flow × potential difference

   = _____ × _____

   Energy transferred = _____ J   [3 marks]

5. **a** Write down the equation that links energy transferred, power and time.

_____ [1 mark]

b A washing machine has a power of 3000 W. Calculate how much energy it transfers when it is switched on for 5 minutes.

Show Me

5 minutes = _____ s

_____

Energy transferred = _____ J   [3 marks]

c Give your answer to part (b) in kilojoules (kJ).

Energy transferred = _____ kJ   [1 mark]

6. A 50 W lamp is switched on for 500 seconds, and a 40 W lamp is switched on for 625 seconds.

Show that both lamps transfer the same amount of energy.

Show Me

50 W lamp: Energy transferred = power × time

= 50 W × _____ s

= _____ J

40 W lamp: _____

_____ [5 marks]

Support
In this question 'show that' means you need to calculate the energy transferred by each lamp. The two amounts of energy transferred should be the same. If they are not, check your working for a mistake. Always show all your working out.

Support
You need to know this equation – it may not be given to you in an exam.

7. A motor uses the mains potential difference. Calculate the charge flow when the motor transfers 500 J of energy.

Show Me

500 J = charge flow × 230 V

Charge flow = ——————————

Charge flow = _____ C   [3 marks]

Support
You need to remember that the mains voltage is about 230 V.

8. Kettle A has a power rating of 2000 W and takes 3 minutes to boil some water. Kettle B has a power rating of 1000 W and takes 6 minutes to boil the same volume of water.

a Explain which kettle transfers the most energy to the water each second.

Show Me

Kettle _____ transfers the most energy each

second because it has a _____

_____.

[2 marks]

Support
Remember that 'explain' means you have to say **what** happens and also **why** it happens.

b Explain which kettle transfers the most energy to the water to bring it to boiling.

Show Me

Both kettles transfer the same amount of energy to boil the same volume of water.

Although kettle B has a lower _____, it is switched on

_____. [2 marks]

c Calculate the amount of energy transferred by kettle A in the 3 minutes it takes to boil the water.

Show Me

3 minutes = 3 × 60 s = _____ s

Energy transferred = _____

Energy transferred = _____ J [3 marks]

d Calculate the charge that flows while kettle A heats the water to boiling. The kettle uses the mains potential difference.

_____

_____

Charge flow = _____ C [3 marks]

Electrical power

- The power of an appliance (in watts, W) is how much energy it transfers each second.
- For an electrical appliance, the power depends on the potential difference and the current:

  power = potential difference × current

- The power can also be calculated from the current and the resistance:

  power = (current)^2 × resistance.

1. State the SI unit for each of the quantities below.

power: _____

current: _____

potential difference: _____

resistance: _____ [4 marks]

2. The potential difference across a bulb is 12 V. The current through the bulb is 4 A.

Calculate the power of the bulb.

Show Me

Power = potential difference × current

= _____ V × _____ A

Power = _____ W [3 marks]

Support
You need to know both of the equations on this page – they may not be given to you in an exam.

3. A 20 W lamp uses the mains voltage of 230 V.

Calculate the size of the current flowing through the lamp.

Show Me

Power = 20 W = 230 V × current

$$\text{Current} = \frac{20\ W}{230\ V}$$

Current = _____ A   [3 marks]

4. A motor has a resistance of 25 Ω. The current in the motor is 10 A. Calculate the power of the motor.

Support
The question gives you a resistance value and a current value. You need to use the equation that links power to resistance and current.

Show Me

Power = (current)² × resistance

= (_____ A)² × _____ Ω

Power = _____ W   [3 marks]

5. A 40 W lamp has a current of 0.4 A flowing through it.

a Calculate the potential difference across the lamp.

Show Me

Power = 40 W = potential difference × _____ A

$$\text{Potential difference} = \frac{\underline{\hspace{2cm}} W}{\underline{\hspace{2cm}} A}$$

Potential difference = _____ V   [3 marks]

b Calculate the resistance of the bulb.

Show Me

Power = 40 W = (0.4 A)² × resistance

$$\text{Resistance} = \frac{40\ W}{\underline{\hspace{2cm}}}$$

Resistance = _____ Ω   [3 marks]

The national grid

- Electrical power is transferred from power stations to users by the national grid. This is a system of cables and transformers.
- Transformers change the potential difference of the electricity. Step-up transformers increase the potential difference and step-down transformers reduce the potential difference.
- Electricity is sent along transmission cables at high voltages and low currents, because then less energy is wasted by heating.
- The potential difference is reduced before the electricity goes into homes to make it safer.

---

1. State whether a **step-down** or a **step-up** transformer is needed for each of the changes below.

> **Support**
> Remember that a 'k' in front of a unit means 'kilo', so 25 kV = 25 000 V.

a Electricity comes out of the power station at 25 kV and is converted to 400 kV to be sent along the transmission lines. _____

[1 mark]

b Electricity from transmission lines at 400 kV is converted to 33 kV to be sent to factories.

_____

[1 mark]

c Electricity from transmission lines at 400 kV is converted to 230 V to be sent into homes.

_____

[1 mark]

2. The diagram shows some parts of the national grid.

Label the diagram using words from the box.

| homes   power station   step-down transformer   step-up transformer   transmission lines |

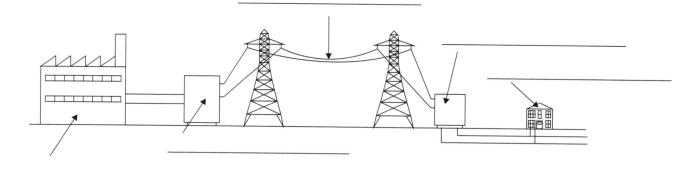

[5 marks]

3. **a** State why electricity is sent along transmission lines at high voltages.

_____

[1 mark]

b State why electricity is reduced to 230 V to be used in homes.

_____

[1 mark]

4. Write these voltages in volts (V).

a 33 kV = _____ V   **b** 400 kV = _____ V   [2 marks]

Matter and density

- Matter is made up of particles.
- The particles are arranged differently in solids, liquids and gases.
- The particle model can be used to explain the differences in density of solids, liquids and gases.

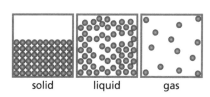

solid   liquid   gas

. . . .

1.   The equation for calculating density is:

$$\text{density} = \frac{\text{mass}}{\text{volume}}$$

Write down the correct SI unit for each quantity in the equation.

mass: _____

volume: *cubic metres m³* _____

density: _____   [3 marks]

2.   The diagrams show the particle arrangements in the different states of matter.

Label the diagrams using words from the box.

solid   liquid   gas   most dense   least dense

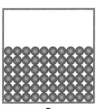

A   B   C

_____   _____   _____

                             _____   _____

[5 marks]

3.   Iron is a metal. Iron can be a solid, a liquid or a gas.

Explain why iron has the greatest density when it is a solid.

The particles are closest together when it is solid, so

Support

Density is the mass of a substance divided by its volume. Try to use the words mass and volume in your answer.

_____

_____

_____   [3 marks]

4. The mass of an iceberg is 183 000 kg. The volume of the iceberg is 200 m³.

Calculate the density of the iceberg.

$$density = \frac{mass}{volume}$$

$$= \underline{\hspace{3cm}}$$

Density = _____ kg/m³   [3 marks]

5. The volume of a brick is 0.002 m³. The density of the brick is 2000 kg/m³.

Calculate the mass of the brick.

$$2000 \text{ kg/m}^3 = \frac{mass \text{ in kg}}{0.002 \text{ m}^3}$$

mass = _____ × _____

Mass = _____ kg   [3 marks]

6. A student used the apparatus shown to find the volume of an object.

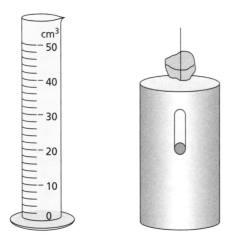

Explain how the student should use this apparatus.

Fill the displacement can with water until it comes up to the spout.

_____

_____

_____

_____   [3 marks]

Changes of state

- Heating increases the energy of particles in the system.
- Heating increases the temperature or causes a change of state.
- The mass of a substance does not change when it changes state.
- Changes of state are physical changes (not chemical changes).
- The temperature of a substance stays the same while it is changing state.

1. Write down the change of state that each of these words describes.

Show Me

a sublimation solid → gas

b evaporation _____

c condensing _____ [3 marks]

2. Give the name used to describe each of these changes of state.

a solid → liquid _____

b liquid → solid _____ [2 marks]

3. What happens to the number of particles in a substance when it changes state?

Tick **one** box.

☐ the number increases   ☐ the number stays the same   ☐ the number decreases   [1 mark]

4. The statements in the table are about physical changes and chemical changes.

Complete the table to show if each statement applies to a **physical change**, a **chemical change** or **both**.

Put **one or two** ticks in each row.

		Physical change	Chemical change
a	The change does not produce a new substance.	☐	☐
b	A new substance is formed.	☐	☐
c	The number of particles is the same before and after the change.	☐	☐
d	The substance has its original properties if the change is reversed.	☐	☐

[4 marks]

5. What is the energy stored within a system called?

Tick **one** box.

☐ internal energy   ☐ chemical energy   ☐ heat energy   [1 mark]

6. A student heats a solid waxy substance in a beaker until all the wax has melted. She measures the temperature every 30 seconds.

The graph shows her results.

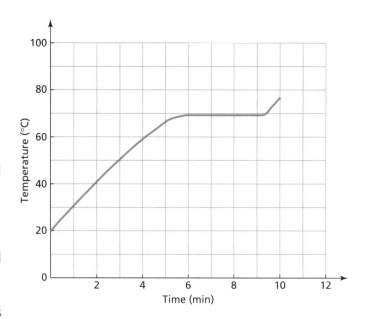

a Draw an X on the graph to show when the substance is melting. [1 mark]

b State the melting point of this substance.

Melting point = _____ °C
[1 mark]

c The student turns off the heat and allows the substance to cool down. She measures the temperature every 30 seconds.

Sketch a graph on these axes to show the change in temperature as the substance cools down.

Support

When you are asked to sketch a graph, it is the shape of the graph that is important. A sketch graph does not usually need numbers on the axes.

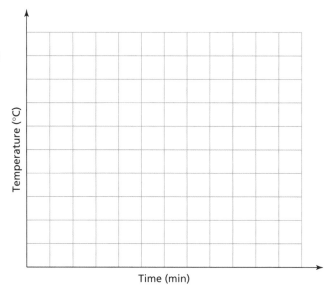

[2 marks]

Latent heat

- The energy needed for a substance to change state is called latent heat.
- The temperature of a substance stays the same while it is changing state.
- Energy supplied while a substance is melting or boiling changes the internal energy of the substance.
- The specific latent heat is the energy needed to change the state of 1 kg of a particular substance.

1. A student is heating water in a beaker. When the water reaches 100 °C he notices that the temperature stops increasing.

a State what happens to water at 100 °C.

_____ [1 mark]

Support

To explain **why**, you need to link what happens (the temperature stops rising) to a **cause**.

Key science words for this topic are internal energy, liquid, gas, state.

b Explain why the temperature stops rising at 100°C, even though energy is still being transferred to the water.

Show Me

The energy transferred to the water is increasing the _____

_____

_____ [2 marks]

2. The energy needed to change 1 kg of a **solid to liquid** is not the same as the energy needed to change 1 kg of the **liquid to gas**.

Write down the change of state that the following terms apply to.

Choose from **solid to liquid** and **liquid to gas**.

a latent heat of vaporisation: _____

b latent heat of fusion: _____ [2 marks]

3. A student melts 500 g of ice. The specific latent heat of fusion for ice is 334 kJ/kg.

Calculate the amount of energy needed to melt the ice.

Use the following equation.

energy needed for a change of state = mass × specific latent heat

Support

This question gives the mass in grams, so you first need to change the mass into kilograms. To change grams to kilograms, divide by 1000.

Show Me

$$\text{Mass in kg} = \frac{500\,g}{1000}$$

Mass = 0.5 kg

Energy needed = _____ kg × _____ kJ/kg

Energy needed = _____ kJ   [3 marks]

Support

The value for specific latent heat is given in units of kilojoules per kilogram, so the energy you calculate using the formula will be in kilojoules.

4. Water in a bowl has a temperature of 15 °C. The bowl of water is put into a freezer. The temperature inside the freezer is −18 °C.

Explain how the temperature of the water changes.

Show Me

At first the temperature _____

This is because _____

While ice is forming _____

Support

There are 4 marks here so you need to give four points. Do not forget to say what happens to the temperature once all the water has frozen.

_____

_____

_____ [4 marks]

Gas pressure

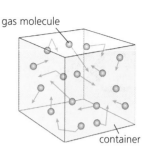
gas molecule

container

- The temperature of a gas is related to the average kinetic energy of the molecules.
- The pressure of a gas increases when the temperature increases, if the gas is kept at a constant volume.
- The particles in a gas collide with the walls of their container. The force from all these collisions is the gas pressure.

1. The statements in the table describe how gas pressure depends on temperature. Tick the boxes to show which statements are correct.

Tick **one** box in each row.

		Correct	Incorrect
a	Particles exert a force when they hit the walls of a container.	☐	☐
b	Particles move faster when they have more kinetic energy.	☐	☐
c	When particles are moving faster they hit the walls of the container more often.	☐	☐
d	When the temperature goes down the kinetic energy of the particles increases.	☐	☐
e	The force on the container walls is greater if the particles are moving more slowly.	☐	☐

[5 marks]

2. Write corrected versions of the statements in question 1 that are wrong.

_____

_____ [2 marks]

3. A gas in a sealed container is cooled down.

Explain why the pressure of the gas gets lower.

Show Me

When the temperature goes down the particles

_____ so they

_____ and hit

_____ less often

Support
In this question you can look at the **correct** statements in question 1 to help you. Arrange them in a sensible order to answer the question. Remember that this question is asking you what happens when the temperature **goes down**.

_____

_____ [4 marks]

Protons, neutrons and electrons

- An atom has a nucleus with a positive charge.
- An atom has no overall electric charge.
- Protons and neutrons make up the nucleus.
- Protons have a positive charge but neutrons have no charge.
- Electrons have a negative charge.

1. Complete the sentences below using words from the box. Use each word only once.

| electrons   neutrons   protons |

Atoms contain positively charged particles called _____. [1 mark]

Most atoms also contain uncharged particles called _____. [1 mark]

There are negatively charged particles called _____ surrounding the nucleus of an atom. [1 mark]

2. Which subatomic particle is **not** found in the nucleus?

Tick **one** box.

☐ proton

☐ electron

☐ neutron   [1 mark]

Support
'Proton' starts with **p**. This helps you remember that **p**rotons are **p**ositive. 'Neutron' starts with **neu** which helps you remember that **neu**trons are **neu**tral. The other particle is the electron which must be negative.

3. The statement below about atoms is wrong.

'A neutral atom has the same number of protons and neutrons.'

Complete the sentence below to give a correct statement.

A neutral atom has the same number of _____

and _____.   [2 marks]

4. Explain why atoms have no overall electric charge.

_____

_____

_____   [2 marks]

Support
Remember that when you **explain why** you should include joining words like 'because' to link your ideas in a sentence.

The size of atoms

- Atoms are very small, with a radius of about 0.1 nanometres (nm).
- 1 nanometre is the same as 10^{-9} m.

$$10^{-9} = \frac{1}{10^9} = \frac{1}{(10 \times 10 \times 10 \times 10 \times 10 \times 10 \times 10 \times 10 \times 10)}$$

- Atoms are mostly empty space. The radius of the nucleus is about 10 000 times smaller than the radius of the whole atom.

---

1. The radius of an atom is about 1×10^{-10} m.

Which of the measurements below is another way of writing 1×10^{-10} m?

Tick **one** box.

☐ 0.1 nanometre

☐ 1.0 nanometre

☐ 10 nanometres   [1 mark]

Support

A negative power of 10 means a number less than 1. The more negative the power of 10, the smaller the number. 10^{-10} m is 10 times smaller than 10^{-9} m.

2. Convert 2.6×10^{-9} m to nanometres (nm).

2.6×10^{-9} m = _____ nm   [1 mark]

3. An atom has a diameter of about 2×10^{-10} m. About how many atoms would fit in 1 nanometre?

Complete the following steps.

> **Show Me**
>
> Convert m to nm: diameter of one atom, 2×10^{-10} m = 0.2 nm
>
> Number of atoms in 1 nm = _____   [2 marks]

Support

Check your answer is sensible. 1 nanometre is not much bigger than the size of one atom, so if you get a very large answer you have made a mistake.

4. The radius of an atom is about 1×10^{-10} m. The radius of a nucleus is about 1×10^{-14} m.

Approximately how many times bigger is an atom compared to its nucleus?

Tick **one** box.

☐ 1000 times bigger

☐ 10 000 times bigger

☐ 100 000 times bigger   [1 mark]

Support

Use this pattern:

10^{-11} is 10 times smaller than 10^{-10}.

10^{-12} is 10×10 times smaller than 10^{-10}.

10^{-13} is $10 \times 10 \times 10$ times smaller than 10^{-10}.

So 10^{-14} is $10 \times 10 \times 10 \times 10 = 10\,000$ times smaller than 10^{-10}.

Elements and isotopes

- Elements are made of one type of atom.
- All atoms of the same element have the same number of protons.
- The number of protons in the nucleus of an atom is called the atomic number.
- The number of protons and neutrons in the nucleus of an atom is called the mass number.
- Atoms of an element can have different numbers of neutrons.
- Atoms of an element with different numbers of neutrons are called isotopes.
- An isotope of carbon (symbol C, atomic number 6) with a mass number of 12 can be shown as $^{12}_{6}C$.

Two isotopes of carbon

1. One isotope of oxygen has a mass number of 16.

   The atomic number of oxygen is 8.

   State the number of neutrons in the nucleus of this isotope.

   _____   [1 mark]

2. Another isotope of oxygen can be represented as $^{18}_{8}O$.

   a Write down the number of protons in the nucleus of this isotope.

 8   [1 mark]

   b Write down the number of neutrons in the nucleus of this isotope. _____   [1 mark]

3. An isotope of sodium (symbol Na) has 14 neutrons and 11 protons in each of its atoms.

   a Write down the atomic number of sodium.

   _____   [1 mark]

   b Write down the mass number of this isotope of sodium.

   _____   [1 mark]

   c Write in symbol form an atom of this isotope. Use the symbol for sodium and the two numbers from your answers to parts (a) and (b).

   _____   [1 mark]

4. Lithium has an atomic number of 3. The diagram shows the nucleus of one atom of an isotope of lithium.

   a State the number of protons in this lithium nucleus.

   _____   [1 mark]

   b State the number of neutrons in this lithium nucleus.

   _____   [1 mark]

Electrons and ions

- When an atom gains or loses electrons it becomes ionised.
- Atoms turn into positive ions if they lose one or more outer electrons.

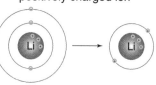

A neutral lithium atom loses an electron and becomes a positively charged ion

1. If an atom loses an electron, what is formed?

   Tick **one** box.

   ☐ A negative ion

   ☐ A positive ion

   ☐ An isotope

   [1 mark]

2. Describe how an atom becomes a positive ion.

   _____   [1 mark]

3. The mass number of sodium is 23 and the atomic number is 11. An atom of sodium loses an electron to form a sodium ion.

   Give the numbers of protons, neutrons and electrons in the sodium ion.

   Number of protons = _____

   Number of neutrons = _____

   Number of electrons = _____   [3 marks]

4. Explain why an atom becomes a positive ion when it loses an electron.

   Show Me

   Atoms have equal numbers of positive and negative charges so an atom is neutral.

   Electrons are _____.

   Therefore, if the atom loses an electron it will have

   more _____.   [3 marks]

Support
For 3 marks you need to make three points and link your sentences together in a sensible order. Useful linking words and phrases include 'because', 'therefore' and 'this means'.

Discovering the structure of the atom

- Before the discovery of electrons, people thought that atoms could not be divided up into smaller particles.
- The discovery of electrons led to the 'plum pudding' model of the atom.
- Alpha scattering experiments by Geiger and Marsden showed that the mass of an atom was concentrated at the centre (the nucleus) and that the nucleus was charged.
- Work by Bohr and Chadwick led to the idea of protons and neutrons inside the nucleus, and electrons orbiting the nucleus.

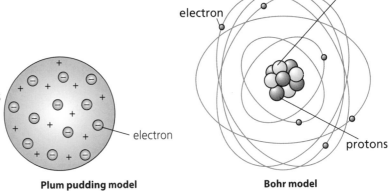

Plum pudding model

Bohr model

1. Complete the sentences below to describe the plum pudding model of the atom.

The 'pudding' part of the atom is a ball of _____ charge.

The 'plums' are particles with a _____ charge.

These particles are called _____.

[3 marks]

2. Draw **one** line from each scientist to the feature of the atom they discovered.

Bohr		The nucleus
Chadwick		Neutrons inside the nucleus
Geiger and Marsden		Electrons at different energy levels

[3 marks]

3. **a** Alpha scattering experiments led to the nuclear model of the structure of the atom, replacing the plum pudding model.

Complete the table to show the correct descriptions of the **nuclear model** and the **plum pudding** model.

_____ **model**	Positive charge spread out everywhere inside
_____ **model**	Positive charge concentrated in a small space

[2 marks]

b In Rutherford's model of the atom, electrons could exist anywhere in a 'cloud' around the nucleus of an atom.

Describe how Bohr's model of the atom is different from Rutherford's model.

_____

_____

[1 mark]

Radioactivity

- Some atomic nuclei are unstable.
- When an unstable nucleus changes to become more stable it gives out radiation. This is radioactive decay.
- Radioactive decay is random.
- The different types of nuclear radiation (radiation from the nucleus) are:
  alpha particle (α) – two neutrons and two protons (the same as a helium nucleus)
  beta particle (β) – a fast-moving electron
  gamma ray (γ) – electromagnetic radiation
  neutron (n).
- Nuclear radiation can be detected by a Geiger-Müller tube.
- The count rate is the number of decays recorded each second.

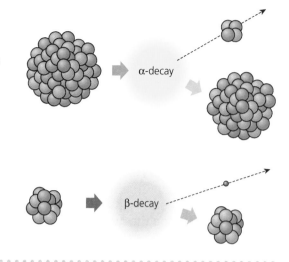

1. Complete this sentence.

   Nuclear radiation comes from the nucleus of an _____ atom.   [1 mark]

2. Complete the table by writing the name of each type of nuclear radiation described.

   Choose from the words in the box. You will not need to use all the words.

   | alpha particle   beta particle   gamma ray   neutron |

   Support
   Look at the words. There is likely to be a link between 'ray' and 'radiation'.

Description	Name of nuclear radiation
An electron	_____
The nucleus of a helium atom (two protons and two neutrons)	_____
Electromagnetic radiation	_____

   [3 marks]

3. Which **two** of the statements below are correct?

   Tick **two** boxes.

   Support
   Look at all the options. Are there any two that cannot both be right?

   ☐ A beta particle is negatively charged.

   ☐ A beta particle is made up of two protons and two neutrons.

   ☐ A beta particle is much smaller than an alpha particle.

   ☐ A beta particle is positively charged.

   [2 marks]

4. A Geiger-Müller tube is used to measure the count rate of a radioactive source that emits alpha radiation.

   Define **count rate**.

   _____   [1 mark]

Comparing alpha, beta and gamma radiation

- Alpha particles are the least penetrating. They are stopped by a few centimetres of air or a few sheets of paper.
- Beta particles travel further in air than alpha particles. They are stopped by a few centimetres of aluminium.
- Gamma rays are the most penetrating.
- Alpha particles are the most ionising (change more atoms into ions).
- Each type of radiation has different uses. The uses are linked to the penetration and the ionising power.

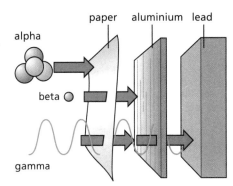

1. Which one of the statements below is true?

   Tick **one** box.

   ☐ Alpha particles are stopped by 5 cm of air.

   ☐ Beta particles are stopped by 5 cm of air.

   ☐ Gamma rays are stopped by 5 cm of air.   [1 mark]

2. Complete the table.

   Use words from the box. Each word may be used more than once.

greatest	least	medium

Radiation type	Distance travelled in air	Ionising power
alpha	_____	_____
beta	_____	medium
gamma	greatest	_____

   [4 marks]

3. Doctors can use gamma radiation to take images of the inside of a patient's body. They put a small amount of radioactive material inside the patient. A radiation detector outside the patient tracks the path of the material through the body.

   Support
   Make sure you understand the word 'emit'. It means give off, or send out.

   Explain why a material that emits **gamma** radiation must be used.

   _____

   _____   [2 marks]

4.

a A scientist investigates the amount of radiation detected at a fixed distance from a radioactive source that emits beta particles.

Suggest how the scientist could investigate how far beta particles can travel through aluminium.

You should include what equipment to use and what measurements to take.

> **Show Me**

First, set up the _____

to measure the count rate when _____ .

Then _____ and measure

_____

Next, change _____

and measure _____ .

Continue increasing _____

until _____ .

[4 marks]

b Give **one** way in which the scientist could make their results more accurate.

_____ [1 mark]

Radioactive decay equations

- A nuclear equation shows the change in mass number and in atomic number during radioactive decay.
- Alpha decay causes the mass of the nucleus to decrease by 4 and the charge of the nucleus to decrease by 2.
- An alpha particle has the symbol $^4_2$He .
- In beta decay a neutron in the nucleus turns into a proton.
- Beta decay does not change the mass of the nucleus, but the charge of the nucleus increases by 1.
- A beta particle has the symbol $^0_{-1}$e.
- Gamma ray emission does not change the mass or the charge of the nucleus.

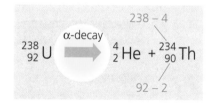

$$^{238}_{92}U \xrightarrow{\alpha\text{-decay}} \,^4_2He + ^{234}_{90}Th$$

238 − 4

92 − 2

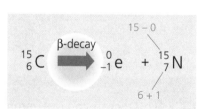

$$^{15}_{6}C \xrightarrow{\beta\text{-decay}} \,^0_{-1}e + ^{15}_{7}N$$

15 − 0

6 + 1

1. Draw **one** line from each of the types of radiation below to the effect that radiation has on the charge of the nucleus.

alpha	Charge decreases by 2
beta	No change
gamma	Charge increases by 1

[3 marks]

2. Which of the statements below describes how a beta particle is produced?

Tick **one** box.

☐ A beta particle is emitted when a proton becomes a neutron.

☐ A beta particle is emitted when a neutron becomes a proton.

☐ A beta particle is emitted when a proton becomes an electron.

☐ A beta particle is emitted when a neutron becomes an electron. [1 mark]

3. Which of the statements below is correct?

Tick **one** box.

☐ The mass of a nucleus increases by 1 when an alpha particle is emitted.

☐ The mass of a nucleus does not change when an alpha particle is emitted.

☐ The mass of a nucleus decreases by 4 when an alpha particle is emitted. [1 mark]

4. An atom of uranium, $^{235}_{92}U$, decays into thorium by emitting an alpha particle.

This reduces the mass of its nucleus by 4 and the charge of its nucleus by 2.

Complete the nuclear equation for this decay.

Support
You need to know the symbol of an alpha particle – it may not be given in an exam.

Show Me

$^{235}_{92}U \rightarrow$ _____ ^{4}He + _____ ^{231}Th [4 marks]

5. An atom of radon, $^{219}_{86}Rn$, decays into polonium by emitting an alpha particle.

Complete the nuclear equation for this decay.

$^{219}_{86}Rn \rightarrow$ _____ He + _____ Po [4 marks]

Support
The total mass number must be the same before and after radioactive decay.

The total atomic number must be the same before and after radioactive decay.

6. An atom of carbon, $^{14}_{6}C$, decays by emitting a beta particle.

Complete the nuclear equation for this decay.

$^{14}_{6}C \rightarrow$ _____ e + _____ N [4 marks]

Half-lives

- The half-life of a radioactive isotope is defined in two ways:
    1. The average time it takes for the count rate (or activity) from a sample containing the isotope to fall to half its initial level.
    2. The average time it takes for the number of nuclei of the isotope in a sample to halve.
- Every radioactive isotope has a particular, constant, value of half-life.

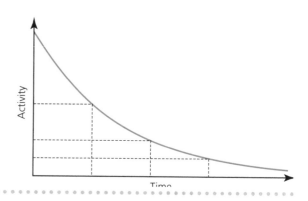

1. Which of the definitions of **half-life** below is correct?

Tick **one** box.

☐ Half the time it takes for the number of nuclei in a sample to halve.

☐ Half the time it takes for all the nuclei in a sample to decay.

☐ The time it takes for the number of nuclei in a sample to halve.

[1 mark]

2. A radioactive material, **X**, has a half-life of 4 hours. A sample of **X** has a mass of 200 g.

Calculate how much of the sample will remain after:

a 4 hours

$$Mass = \frac{200\ g}{2}$$

Support
Always show **all** your working out. If you make a mistake you may still get marks for the working.

Mass = 100 g   [1 mark]

b 8 hours

After another 4 hours (8 hours total), mass $= \dfrac{100\ g}{2}$

Mass = _____ g   [1 mark]

c 12 hours.

_____

Mass = _____ g   [1 mark]

3. The activity of a sample of radioactive material falls from 400 counts per second to 100 counts per second in 2 hours.

Calculate the half-life of this material.

_____

_____

Half-life = _____ hours   [2 marks]

4. The graph shows how the activity of a sample of radioactive material changes over time.

Find the half-life of this radioactive material.

You must draw lines on the graph to show how you calculated your answer.

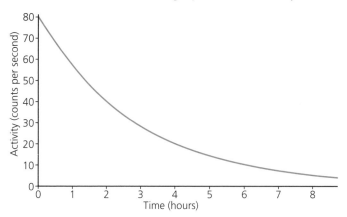

Show Me

The activity halves from _____ to

_____ in _____ hours.

Half-life = _____ hours   [2 marks]

5. The table shows how the activity of a sample of a radioactive material changes over time.

Time (hours)	Activity (counts per minute)
0	60
1	43
2	30
3	20
4	15

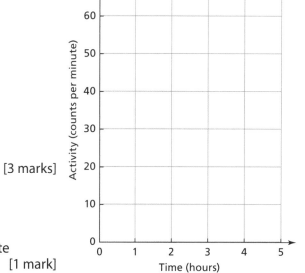

a Plot the results from the table on the graph grid. Draw a line of best fit.   [3 marks]

b Use your graph to find the activity of the sample after 2.5 hours.

Activity = _____ counts per minute   [1 mark]

c Use your graph to find the half-life of this radioactive material.

_____

_____

Half-life = _____ hours   [2 marks]

Support
Remember a best fit line can be curved or straight – you must decide which fits the pattern of your plotted points best.

Radioactive hazards

- Radioactive sources can cause harm to living things.
- Precautions must be taken to protect people from the hazard of a radioactive source.
- The type of radiation emitted (alpha, beta or gamma) affects the level of hazard.
- Radioactive contamination is the unwanted presence of radioactive materials.
- Irradiation means exposing an object to X-rays or radiation from a radioactive source.
- Irradiation harms living things but the irradiated object does not become radioactive.

1. Which one of the statements below is correct? Tick **one** box.

☐ All three types of radioactivity cause the same level of hazard.

☐ Objects irradiated by gamma rays become radioactive.

☐ The level of hazard depends on the type of radioactivity emitted. [1 mark]

2. Explain why alpha particles cause more damage than beta particles and gamma rays if they enter your body.

_____ [1 mark]

3. Food irradiation is a process that exposes food to X-rays or gamma rays. This kills bacteria in the food. A conveyor belt is used to move food past a source of gamma radiation. The room where this happens has thick concrete walls.

a State the purpose of the thick concrete walls.

_____ [1 mark]

b The food is uncontaminated by radioactivity. What does this mean?

Tick **one** box.

☐ The food has not been irradiated. ☐ There is no radioactive material on the food. [1 mark]

4. In schools, radioactive sources are kept in special boxes. These boxes are made from wood about 10 cm thick. There is a layer of lead inside, about 0.5 cm thick, around the radioactive source.

State and explain how well one of these boxes protects you from alpha, beta and gamma radiation from the radioactive source.

Support
You need to say how well you would be protected from **each** of alpha radiation, beta radiation and gamma radiation, and also give a reason for each of your decisions. Useful words and phrases include 'so', 'because', 'but' and 'this means'.

_____

_____

_____

_____ [4 marks]

Forces

- Scalar quantities, like distance, have a magnitude (size) only. They do not have a direction.
- Vector quantities, like force, have a direction as well as a size, and can be represented by arrows.
- Forces can be contact or non-contact forces.
- The effects of two or more forces can be combined into one resultant force.

1. Speed is a scalar quantity.

Name the vector quantity that is speed in a particular direction.

_____ [1 mark]

> **Support**
> Momentum is a combination of mass and velocity, and acceleration is a change in velocity. Often, if something depends on a vector quantity it is a vector quantity itself.

2. Write S or V next to these quantities, to show whether they are scalars or vectors.

Show Me

mass ☐   force ☐   energy ☐   time ☐ S

velocity ☐   momentum ☐ V   temperature ☐   acceleration ☐   [8 marks]

3. **a** Non-contact forces can affect objects without touching them.

Write down the names of **three** non-contact forces.

Show Me

i. *electrostatic force*

ii. _____

iii. _____   [3 marks]

b Air resistance is a contact force.

Write down **two** other examples of contact forces.

i. _____

ii. _____   [2 marks]

4. The diagram shows a 20 N force acting on a box.

☐ →20 N   ☐ A

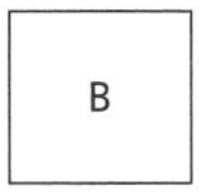

> **Support**
> In maths you may see a vector shown like this: **PQ̄**. In science, a vector is represented by an arrow pointing in the correct direction with the length of the arrow representing the magnitude of the vector.

a Draw an arrow on box A to represent a 40 N force in the same direction as the 20 N force.   [1 mark]

b Draw an arrow on box B to represent a 20 N force in the opposite direction to the forces above.   [1 mark]

☐ B

The drawings show different horizontal forces acting on a car.

Calculate the **resultant** force on each car. State the direction in which it is acting.

a

Support
When two forces are acting in the same direction, find the resultant by adding the sizes of the forces together.

Resultant = 1000 N + 200 N

Resultant = _____ N acting backwards   [2 marks]

b

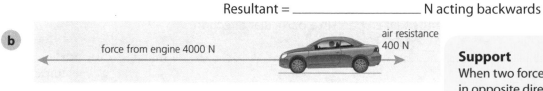

Support
When two forces are acting in opposite directions, find the resultant by subtracting one force from the other. The resultant will be acting in the direction of the largest force.

Resultant = _____ N acting _____   [2 marks]

c

Resultant = _____ N   [1 mark]

d

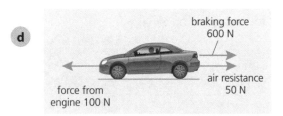

Resultant = _____ N acting _____   [2 marks]

Speed and velocity

- Distance in a particular direction is a vector called displacement.
- Speed is a scalar quantity. Velocity is a vector.
- Speed is calculated from measurements of distance and time:

  distance travelled = speed × time

- The speed of a moving object usually changes. To find the average speed, use the journey time and distance travelled.
- The speed of the wind and the speed of sound can change.

1.   Car A is moving along a straight road at 15 mph. Car B is going around a roundabout at 15 mph.

a   State and explain which car (or cars) is moving at a constant speed.

_____   [1 mark]

b   State and explain which car (or cars) is moving at a constant velocity.

_____   [1 mark]

c Car B travels 50 m around the roundabout once, and ends up where it started.

    i. State the distance that the car travelled.

Distance = _____ m   [1 mark]

    ii. State the displacement at the end of the car's journey around the roundabout.

Displacement = _____ m   [1 mark]

2. Draw one line from each of the left-hand boxes to one of the right-hand boxes, to show typical speeds.

Show Me

walking	330 m/s
cycling	3 m/s
running	6 m/s
speed of sound	10 m/s
car in town	40 m/s
train	1.5 m/s

[6 marks]

3. An athlete runs at 5 metres per second for 5 minutes. Calculate the distance she runs.

Show Me

5 minutes = 5 × 60 s = _____ s

Distance = speed × time

= _____ m/s × _____ s

Distance = _____ m   [4 marks]

4. A cyclist takes 15 minutes to travel 4500 metres. Calculate his average speed in m/s.

Show Me

15 minutes = _____

Distance = 4500 m = average speed in m/s × _____ s

Average speed = $\dfrac{4500 \, m}{s}$

Average speed = _____ m/s   [4 marks]

5. A car is travelling at 20 m/s. Calculate how long it takes to cover 10 km.

Show Me

10 km = _____ m

_____

_____

_____

Time = _____ s   [4 marks]

6. Describe how you could find the speed of a student walking across the playground.

Show Me

I will use a tape measure to measure a distance of _____.

_____

_____

I will calculate the speed by _____.   [3 marks]

Acceleration

- Acceleration is a change in speed in a certain time.
- Acceleration is a vector quantity.
- An object that is slowing down is decelerating. Its acceleration is negative.
- Acceleration, change in velocity and time are linked by this equation:

$$\text{acceleration} = \frac{\text{change in velocity}}{\text{time taken}}$$

1. Write down the correct unit for each of these quantities.

acceleration: _____

change in velocity: _____

time: _____   [3 marks]

Support
Acceleration is the rate of change of velocity. This means **how much the velocity changes** each second. The units are **metres per second per second**, which is written as m/s^2.

2. A car increases its velocity by 10 m/s in 5 s. Calculate the car's acceleration.

Show Me

$$\text{Acceleration} = \frac{\text{change in velocity}}{\text{time taken}}$$

$$= \underline{\qquad}$$

Support
You need to know this equation – it may not be given to you in an exam.

Acceleration = _____ m/s^2 [3 marks]

3. A car changes its velocity from 20 m/s to 5 m/s in 5 seconds. Calculate its acceleration.

Change in velocity = final velocity – initial velocity

$$= 5 \text{ m/s} - 20 \text{ m/s}$$

$$= -15 \text{ m/s}$$

Acceleration = _____

Support
The minus sign for the change in velocity shows that the car has slowed down. Make sure you use this value when you calculate the acceleration. Your final answer should also have a minus sign.

Acceleration = _____ m/s² [3 marks]

4. A car travelling at 20 m/s brakes and comes to a stop. Its acceleration is –5 m/s².

Calculate the time it takes for the car to come to a stop.

Change in velocity = 0 – 20 m/s = –20 m/s

$$\text{Acceleration} = -5 \text{ m/s}^2 = \frac{-20 \text{ m/s}}{\text{time in s}}$$

Time = _____

Time = _____ s [4 marks]

Motion graphs

- Journeys can be represented on distance–time graphs or velocity–time graphs.
- A sloping line on a distance–time graph represents a constant speed. A horizontal line represents zero speed.
- The gradient (slope) of the line on a distance–time graph gives the speed.
- A sloping line on a velocity–time graph represents acceleration. A horizontal line represents constant speed.
- The gradient of the line on a velocity–time graph gives the acceleration.

1. Look at the graph on the right.

Write down how far the person had jogged after 10 seconds.

Distance = _____ m [1 mark]

2. Look at the graph. Write down the **start time** and the **end time** for the period when

a the jogger was stationary:

_____ [1 mark]

b the jogger was moving fastest:

_____ [1 mark]

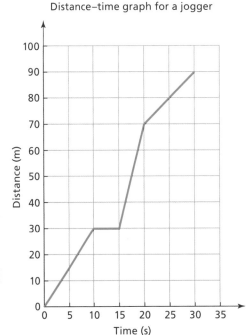

Distance–time graph for a jogger

3. Look at the graph on the previous page. Calculate the speed of the jogger between 15 and 20 seconds.

Show Me

$$Speed = gradient = \frac{change\ in\ distance\ (vertical\ distance\ on\ graph)}{change\ in\ time\ (horizontal\ distance\ on\ graph)}$$

$$= \frac{(70\ m - 30\ m)}{(20\ s - 15\ s)} = \underline{\hspace{3cm}}$$

Speed = _____ m/s   [3 marks]

4. Look at the graph on the previous page. Calculate the speed of the jogger between 20 and 30 seconds.

_____

_____

_____

Speed = _____ m/s   [3 marks]

5. The table shows data for another runner.

Plot the data on the graph grid. Join the points to show this runner's journey.

Time (s)	Distance (m)
0	0
5	10
20	55

[2 marks]

6. The **velocity–time graph** below shows the journey of a cyclist.

State the cyclist's velocity at 20 seconds from the start of the journey.

Velocity = _____ m/s   [1 mark]

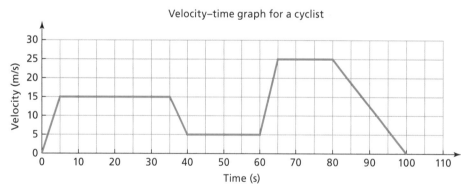

Velocity–time graph for a cyclist

7. Look at the graph in question 6. Write down the start time and end time for **one** period when the cyclist was:

    a speeding up: _____ [1 mark]

    b slowing down: _____ [1 mark]

    c travelling at a constant speed: _____ [1 mark]

8. Look at the graph in question 6. Calculate the acceleration of the cyclist between 60 and 65 seconds. Give the unit with your answer.

Change in velocity = 25 m/s – 5 m/s = 20 m/s

Change in time = 65 s – 60 s = 5 s

Acceleration = gradient = $\dfrac{\text{change in velocity}}{\text{change in time}}$ = _____

                Acceleration = _____ Unit _____ [4 marks]

9.   **a** Look at the graph in question 6. Calculate the acceleration of the cyclist between 80 and 100 seconds.

Change in velocity = 0 m/s – _____

Change in time = 100 s – _____

Acceleration = _____

> **Support**
>
> Your answer to part (a) should have been a negative number. Think about what a negative acceleration means.

                Acceleration = _____ m/s² [2 marks]

    b Describe what is happening to the velocity of the cyclist during this time.

_____ [1 mark]

10. The table shows how the velocity of another cyclist changes during the ride.

Plot the data on the graph above. Join the points to show this cyclist's journey.

Time (s)	Velocity (m/s)
0	0
5	10
50	10
55	15
100	15

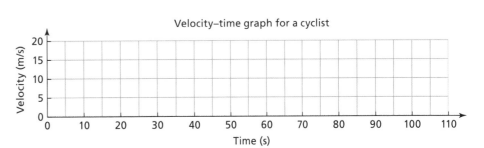

[2 marks]

Acceleration and distance

- An acceleration can be linked to the distance over which it happens by the following equation:

  (final velocity)^2 – (initial velocity)^2 = 2 × acceleration × distance

Support
You will also see this equation written in symbols as $v^2 - u^2 = 2as$. You can see that a represents acceleration, but you will have to remember that u represents initial velocity, v represents final velocity and s represents distance.

1. Write down the correct unit for each of the quantities in the equation.

Show Me

final velocity: m/s initial velocity: ______________

acceleration: ______________ distance: ______________ [4 marks]

2. A car accelerates from 5 m/s to 10 m/s. It covers a distance of 90 m while it is accelerating.

Calculate the acceleration of the car.

Show Me

Final velocity = 10 m/s, initial velocity = 5 m/s

(final velocity)^2 – (initial velocity)^2 = 2 × acceleration × distance

(10 m/s)^2 – (5 m/s)^2 = 2 × acceleration × 90 m

100 – 25 = 2 × 90 × acceleration

______________ = ______________ × acceleration

Acceleration = ______

Support
You do not need to remember this equation, but you will need to choose it from a list of equations in an exam.

Acceleration = ______________ m/s^2 [3 marks]

3. A car decelerates from 25 m/s to 10 m/s. Its acceleration is –6 m/s^2.

Calculate how far the car travels while it is slowing down.

Show Me

Final velocity = ______________ m/s,

initial velocity = ______________ m/s

____

____

____

Support
Remember that a negative acceleration means the object is slowing down.

Distance = ______________ m [3 marks]

4. A train sets off from a station. It travels for 1 km while it accelerates to a velocity of 40 m/s.

Calculate its acceleration.

_____

_____

_____

_____

Acceleration = _____ m/s^2 [4 marks]

Forces and acceleration

- Weight is the force on an object due to gravity. It is caused by the gravitational field of the Earth.
- Weight is directly proportional to mass. The equation is:

  weight = mass × gravitational field strength (g)

  To use the equation, weight must be in newtons and mass in kilograms.
- The Earth's gravitational field strength (g) is 10 N/kg (newtons per kilogram).
- We assume the weight of an object acts at a point called the centre of mass.
- If there is no resultant force, a stationary object remains stationary and a moving object continues to move at the same velocity. This is Newton's first law.
- The equation linking force, mass and acceleration is:

  force = mass × acceleration

- The equation can be rearranged to give:

  $$\text{acceleration} = \frac{\text{force}}{\text{mass}}$$

  This shows that the acceleration of an object is proportional to the resultant force, and inversely proportional to the mass. This is Newton's second law.

1. A box of shopping has a mass of 15 kg. The gravitational field strength on Earth is 10 N/kg.

Calculate the weight of the box. Give the unit of weight.

> **Show Me** Weight = mass × gravitational field strength

Support
You need to know this equation – it may not be given to you in an exam.

= _____

Weight = _____ Unit _____   [4 marks]

2. The weight of a bicycle is 90 N. The gravitational field strength on Earth is 10 N/kg.

Calculate the mass of the bicycle.

> **Show Me** Weight = 90 N = mass × 10 N/kg

$$\text{Mass} = \frac{90\,N}{\underline{\quad\quad}}$$

Mass = _____ kg   [3 marks]

3. A car is moving at 20 m/s. The force from the engine is 300 N. The air resistance and friction forces total 300 N.

Explain what will happen to the speed of the car.

Support
If a question asks you explain, you need to say **what** happens and also **why** it happens.

_____

_____ [2 marks]

4. A girl on a bicycle is pedalling using a force of 25 N. She is speeding up.

The air resistance and friction forces must total less than 25 N. Explain how you know this.

Show Me

If she is speeding up, the resultant force on her _____

so the backwards force must be _____. [2 marks]

5. Which of these expressions is correct? Tick **one** box.

Support
the symbol ∝ means 'proportional to'.

☐ force ∝ mass

☐ force ∝ acceleration

☐ acceleration ∝ mass [1 mark]

6. The acceleration on an object depends on its mass and on the resultant force.

Which of these statements are true? Tick **two** boxes.

☐ If the force doubles, the acceleration doubles.

☐ If the mass doubles, the acceleration doubles.

☐ If the force is only half as big, the acceleration doubles.

☐ If the mass is only half as big, the acceleration doubles. [2 marks]

7. A motorbike has a mass of 200 kg. It can accelerate at 9 m/s².

a Calculate the force needed from the engine to produce this acceleration.

Support
You need to know this equation – it may not be given to you in an exam.

Show Me

Force = mass × acceleration

_____

Force = _____ N [3 marks]

b The same motorbike accelerates with an engine force of 1000 N.

Calculate the acceleration.

_____

_____

_____

Acceleration = _____ m/s² [3 marks]

8. Two students are investigating how the mass of a trolley affects its acceleration. They use the same force to accelerate the trolley each time. They put different masses on the trolley.

a Describe how the students could measure the acceleration of the trolley.

Show Me

They could use light gates to measure the speed of the trolley _____

_____ and the time it takes the

trolley to _____ .

Then they calculate the _____

_____ [3 marks]

b The students plot a graph to show their results. Which graph here shows what their results should look like?

Tick **one** box.

A acceleration / mass   B acceleration / mass   C acceleration / mass   D acceleration / mass

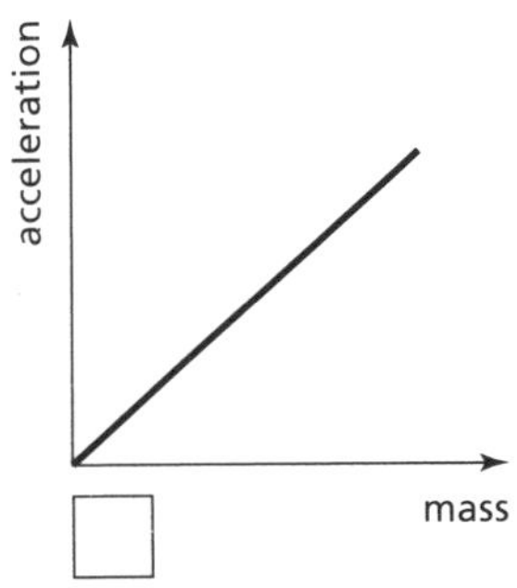

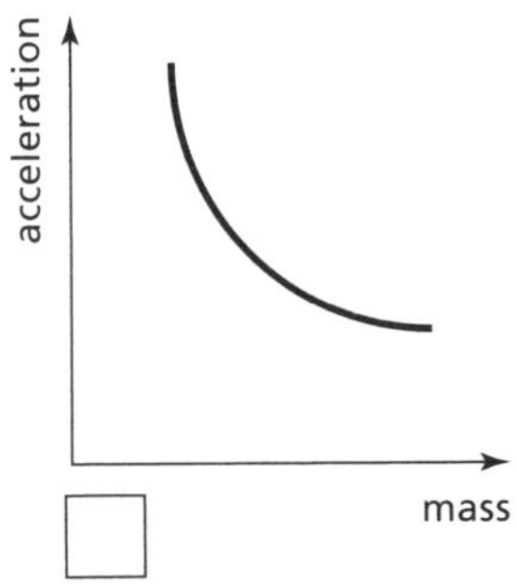

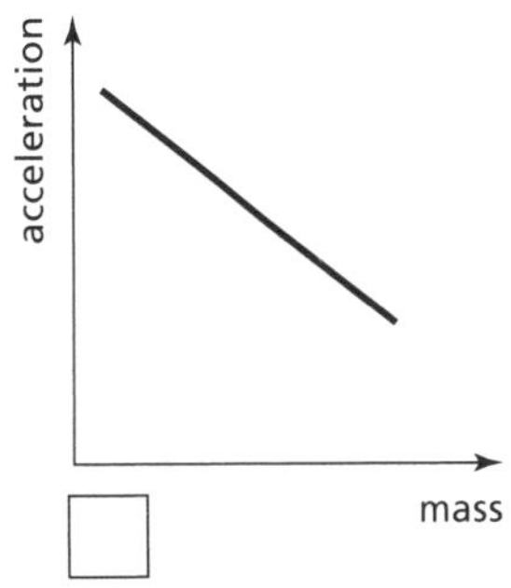

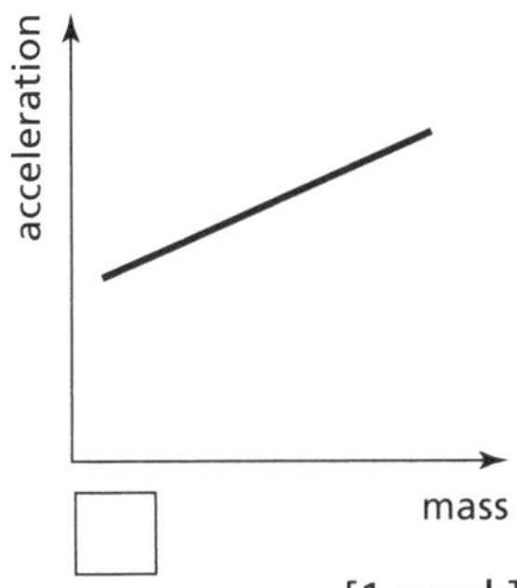

[1 mark]

c A data logger measures the speed of a trolley by timing how long it takes the card on the trolley to pass between light gates. The students enter an incorrect value for the length of the card into the data logger.

Will this cause a **random** or a **systematic** error? Explain your answer.

_____

_____ [1 mark]

Free fall

- Objects fall with an acceleration of about 10 m/s².
- A falling object accelerates due to the force of gravity.
- As the falling object speeds up, the air resistance (or water resistance) increases.
- Eventually the resultant force is zero and the object stops accelerating.
- The constant speed that a falling object reaches is called its terminal velocity.

1. The drawings show a skydiver at different stages in a jump.

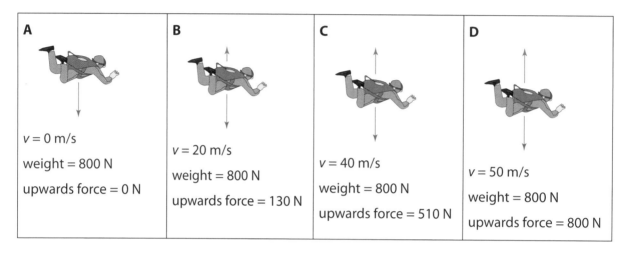

A	B	C	D
v = 0 m/s weight = 800 N upwards force = 0 N	v = 20 m/s weight = 800 N upwards force = 130 N	v = 40 m/s weight = 800 N upwards force = 510 N	v = 50 m/s weight = 800 N upwards force = 800 N

a What causes the upwards force on the skydiver?

_____ [1 mark]

b Describe how the size of this force changes as the skydiver's speed changes.

_____ [1 mark]

2. State the size of the resultant force on the skydiver in drawing A above.

Show Me

_____ N in a _____ direction. [1 mark]

3. The acceleration of an object depends on its mass and on the resultant force.

Explain how the acceleration of the skydiver changes as she falls. Refer in your answer to the diagrams A to D.

Show Me

The mass of the skydiver _____

so the acceleration depends only on the _____.

The _____ gets _____

as she falls, so _____ [4 marks]

Action and reaction forces

- A pair of objects can exert forces on each other.
- The forces between objects in a pair are equal and opposite. These are sometimes called action and reaction forces.
- Action and reaction forces are always the same size, in opposite directions and act on *different* objects.
- Action and reaction forces are not the same as balanced forces. Balanced forces act on the *same* object and produce a zero resultant force.

1. All objects with mass attract each other. The force of attraction is called gravity.

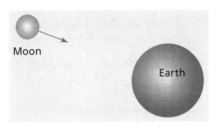

The diagram shows the force on the Moon caused by the Earth's gravity.

Draw an arrow on the Earth to show the force on the Earth caused by the Moon's gravity. [2 marks]

2. Diagram **a** and diagram **b** each show one of the forces between a pair of magnets.

Draw an arrow on the other magnet in each pair to show the force on the other magnet.

Draw **one** arrow in diagram **a** and **one** arrow in diagram **b**. [2 marks]

> **Support**
> Remember that forces are vectors. The direction of the arrow shows the direction of the force. The length of the arrow shows the size of the force.

a | S   N ← | N   S |

b | S   N → | N   S |

3. A boy is standing on the floor. His weight is 400 N. This means he is pushing on the floor with a force of 400 N. There is a force from the floor pushing up on him, called the normal contact force.

State the size of the normal contact force.

_____ N [1 mark]

4. Some of these sentences describe balanced forces. Some describe action and reaction pairs of forces.

Tick **one** box for each line.

		Action and reaction pair	Balanced forces
a	The force from the engine of a boat is the same size and in the opposite direction to the water resistance on the boat.	☐	☐
b	The weight of the boat pushes down on the water, and upthrust from the water pushes up on the boat.	☐	☐
c	Gravity pulls the boy towards the Earth, and the normal contact force from the floor pushes up on the boy.	☐	☐
d	A dog on a lead pulls on its owner, and the owner pulls on the dog.	☐	☐

[4 marks]

Forces and braking

- The friction force in brakes does work. It transfers kinetic energy stored in the moving vehicle into thermal energy which heats up the brakes.
- Faster vehicles need a greater braking force to stop in a certain distance.
- The greater the braking force, the greater the deceleration of the vehicle.
- Friction caused by a large deceleration could make the brakes overheat.
- A large deceleration could also cause the driver to lose control and the vehicle could skid.

1. Choose words from the box to complete the sentences. You will not need to use all the words.

acceleration	distance	friction	greater	kinetic	reduced	smaller	thermal

When a vehicle brakes, _____ causes the _____ energy of the vehicle

to be _____. The greater the speed of the vehicle, the _____ the

braking force needed to stop it in a certain _____. [5 marks]

2. A large deceleration requires a large braking force. Describe **two** possible effects of large braking forces.

Show Me

Energy transferred by braking may make the brakes _____.

Large forces may make the driver _____. [2 marks]

3. **a** Estimate the speed of a car driving in a town.

> **Support**
> The symbol ≈ stands for 'approximately equal to'.
>
> You need to remember some typical speeds. Question 2 on page 49 may help you.

Show Me

Speed ≈ 10 m/s [1 mark]

b A car starts off from traffic lights and accelerates to the speed you gave in part (a). This takes about 5 seconds.

i. Write down the equation that links change in speed, acceleration and time.

_____ [1 mark]

ii. Estimate the car's acceleration.

_____ [1 mark]

Acceleration = _____ m/s² [2 marks]

c **i.** Write down the equation that links force, mass and acceleration.

_____ [1 mark]

ii. The car has a mass of 1500 kg. Estimate the force that caused the acceleration you calculated in part (b).

_____

Force ≈ _____ N [2 marks]

Stopping distances

- The stopping distance of a vehicle is how far it moves, from the time a driver sees a hazard to when the vehicle stops.
- The stopping distance is made up of the thinking distance and the braking distance.
- The thinking distance depends on the driver's reaction time. This is affected by factors to do with the driver, such as tiredness.
- The braking distance depends on factors to do with the car or the road, such as worn tyres or an icy road.

1. Reaction times vary from person to person. What is a typical value for a person's reaction time?

Tick **one** box.

☐ 0.1 s   ☐ 0.5 s   ☐ 1.0 s   ☐ 5.0 s   [1 mark]

2. Two groups of students measure their reaction times.

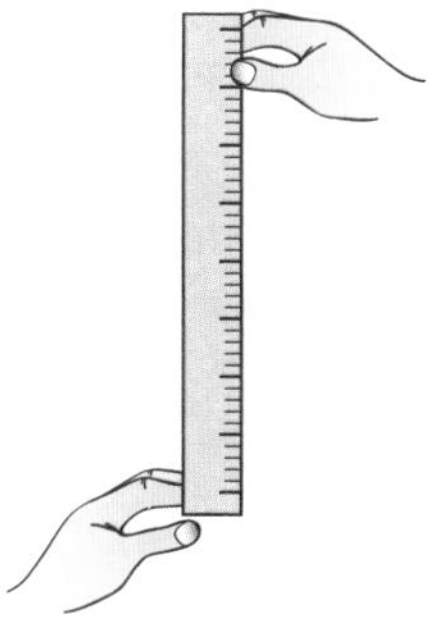

Group **X** uses a ruler and this method:

1 One student holds the top of the ruler as shown here, then lets go.

2 Another student catches it, as shown.

3 They work out the reaction time from how far the ruler falls before it is caught.

Group **Y** uses computer software that shows a view from a car driver's seat. The student has to press a button when they see a dog running into the road.

Evaluate the methods of group **X** and group **Y**.

Show Me

The ruler method can be done with simple apparatus, whereas

the computer method _____.

However the computer method is more realistic because

_____.

I think _____ is better because

_____

_____   [3 marks]

Support
Evaluate a method means to discuss the strengths and weaknesses of the method. Here, you are asked evaluate the **two** methods for finding reaction time, so you should list some advantages and disadvantages of each method. You also need to decide which is best and give a reason for this.

3. A cat runs into the road in front of a car. The car travels 10 m before the driver presses the brake pedal. The car then travels a further 15 m before it comes to a stop.

   a State the thinking distance.

   _____ m   [1 mark]

   b State the braking distance.

   _____ m   [1 mark]

   c Calculate the stopping distance.

   _____

   _____ m   [1 mark]

4. The Highway Code shows standard **thinking distance** and **braking distance** for different speeds of a typical car.

State which distance is affected by each of the following factors, and how it is affected.

Show Me

a Worn brakes *increase the braking distance.* [1 mark]

b A wet road _____. [1 mark]

c A driver using a mobile phone _____. [1 mark]

5. A driver with normal alertness is driving a car at 10 m/s. The driver's thinking distance at this speed is 7 m.

a State how the thinking distance of a driver depends on the car's speed.

Show Me

Thinking distance = reaction time × speed

The reaction time is _____,

so _____

_____ [3 marks]

b State the thinking distance for this driver travelling at 20 m/s. _____ m [1 mark]

c The driver is distracted and her reaction time increases to 1.2 s.

What is her thinking distance when driving at 20 m/s? Tick **one** box.

☐ 14 m   ☐ 16 m   ☐ 24 m   [1 mark]

Support
Do not just guess the answer. Work it out using the information given.

6. Speed limits on roads in the UK are different for different types of roads. For example, the speed limit outside a school may be 20 mph, but the speed limit on most motorways is 70 mph.

Explain why speed limits are different on different types of road.

Your answer should include an explanation of how the speed of a vehicle affects the stopping distance.

Support
This is an extended writing question, worth 6 marks. Your answer needs to include the explanation asked for in the question, and you need to link your sentences together in a sensible order. You could use bullet points or subheadings.

Show Me

The stopping distance of a vehicle is made up of the thinking distance and the braking distance. At greater speeds,

_____

_____

_____

_____

_____

_____ [6 marks]

Force and extension

- Pairs of forces can change the shape of an object by stretching, bending or compressing it.
- Objects can stretch elastically or inelastically.
- The extension or compression of an object such as a spring is proportional to the force applied, as long as the limit of proportionality is not exceeded.
- The equation is:

  force applied to spring = spring constant × extension

- To use the equation, force must be in newtons and extension in metres. The units of the spring constant are N/m (newtons per metre).

---

1. A student pushes a spring.

   a Describe what will happen to the spring.

_____ [1 mark]

   b Explain why you need more than one force to change the shape of an object.

_____ [1 mark]

2. Objects such as springs, rubber bands or modelling clay can change shape elastically or inelastically.

Write an **E** next to the best description of **elastic** deformation. Write **one** E only.

Write an **I** next to the best description of **inelastic** deformation. Write **one** I only.

☐ A graph of force against extension is a straight line.

☐ The object goes back to its original shape when the forces are removed.

☐ The larger the force, the larger the extension.

☐ The extension is proportional to the force.

☐ The object does not go back to its original shape when the forces are removed. [2 marks]

3. A student is investigating how the extension of a spring depends on the force on it. She is using the apparatus shown.

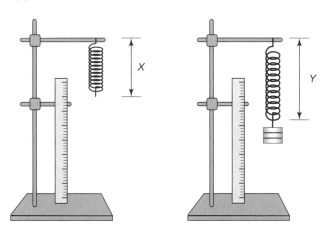

a Look at the diagram on the previous page. Explain why distances X and Y are both needed to work out the extension of the spring.

The extension is how far the spring stretches, so you need to know

_____

_____ [2 marks]

The student adds masses to the spring and measures the length each time. The table shows her results.

Weight on spring (N)	Extension (mm)			
	1st go	2nd go	3rd go	mean
0	0	0	0	0
1	4.8	5.1	5.1	5.0
2	10.2	10.0	9.8	10.0
3	15.3	15.0	14.9	15.1
4	19.9	20.1	20.2	20.1
5	24.9	24.5	25.2	____

b Calculate the missing mean and write it in the table. [1 mark]

c Explain why the student took three measurements for each different weight and then calculated a mean.

Repeating measurements lets you spot any results _____.

_____

_____

_____ [3 marks]

d Plot the student's results on the graph axes. Draw a line of best fit through the points. [3 marks]

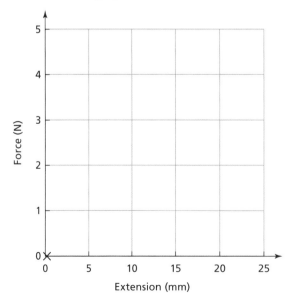

e The student repeats the investigation. This time he fastens a short piece of wire to the bottom of the spring, so that the wire is horizontal and almost touches the ruler.

Explain how this will affect the accuracy of his measurements.

_____

_____ [2 marks]

4. A spring has a spring constant of 20 N/m. The spring is stretched, with an extension of 5 cm.

Calculate the force needed to give this extension.

$5 \text{ cm} = \dfrac{5}{100} = 0.05 \text{ m}$

Force = spring constant × extension

= _____

Force = _____ N   [4 marks]

5. A spring has an extension of 0.3 m when a 6 N force is applied to it.

Calculate the spring constant of the spring. Give the unit.

Force = 6 N = spring constant × 0.3 m

Spring constant = $\dfrac{6 \text{ N}}{\text{____}}$

Spring constant = _____ unit _____ N/m   [4 marks]

6. The graph shows force and extension for two springs. State and explain which spring, **A** or **B**, has the largest spring constant.

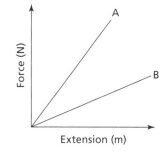

Spring A has the largest spring constant because it takes

_____

_____ [2 marks]

Springs and energy

- A force stretching a spring does work. Elastic potential energy is stored in the stretched spring.

  elastic potential energy = work done to stretch the spring

- The elastic potential energy stored in a spring can be calculated using this equation:

  elastic potential energy = $0.5 \times$ spring constant $\times$ (extension)^2

1. Which of these sentences is the best description of the spring constant of a spring?

   Tick **one** box.

   ☐ The force needed to stretch the spring by 1 m.

   ☐ The extension of the spring when a 1 N force is applied to it.

   ☐ The maximum force the spring can support without breaking. [1 mark]

2. A spring has a spring constant of 15 N/m. It is extended by 20 cm.

   Calculate the elastic potential energy stored in the stretched spring.

   Support
   You do not need to remember this equation – you would be able to find it on an equation sheet in an exam.

   Show Me

   $20 \text{ cm} = \dfrac{20}{100} = 0.2 \text{ m}$

   Elastic potential energy = $0.5 \times$ spring constant $\times$ (extension)^2

   $= 0.5 \times 15 \text{ N/m} \times (0.2 \text{ m})^2$

   Elastic potential energy = _____ J [3 marks]

3. A spring with a spring constant of 100 N/m is 20 cm long. It is stretched until its total length is 50 cm.

   Calculate the elastic potential energy stored in the spring.

   Show Me

   Extension = 50 cm – _____ = _____ cm = _____ m

   _____

   _____

   Elastic potential energy = _____ J [4 marks]

4. A spring is stretched by 0.6 m. It stores 20 J of elastic potential energy.

   Calculate the spring constant.

   Show Me

   Energy stored = 20 J = $0.5 \times$ spring constant $\times$ (0.6 m)^2

   Spring constant = $\dfrac{20 \text{ J}}{0.5 \times (0.6 \text{ m})^2}$

   $= \underline{}$

   Spring constant = _____ N/m [4 marks]

Transverse and longitudinal waves

- Waves are produced by vibrations.
- Transverse waves travel at 90° to the direction of vibration.
- Longitudinal waves travel parallel to the direction of vibration.
- Longitudinal waves have areas of compression (pushed together) and rarefaction (pulled apart).
- The amplitude of a wave is the maximum displacement of a point on the wave (from its undisturbed position).
- The wavelength of a wave is the distance from one point on the wave to the next point that vibrates in the same way.

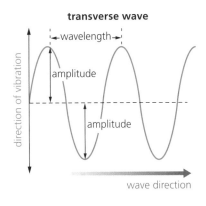

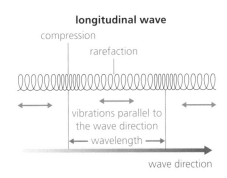

1. Sound travels through air as waves.

State the type of wave that describes how sound waves travel.

_____ [1 mark]

2. Amy throws a stone into a pond. This creates ripples on the water surface. The ripples are transverse waves.

What moves outwards from where the stone is dropped? Tick **one** box.

☐ energy ☐ water ☐ the stone [1 mark]

3. Which one of the statements below is correct? Tick **one** box.

☐ Longitudinal waves travel at right angles to the vibrations that produced the wave.

☐ Longitudinal waves do not have a wavelength.

☐ Longitudinal waves have compressions and rarefactions. [1 mark]

4. Tick **one** box in each row, to show whether the statement is true for **longitudinal waves only**, **transverse waves only**, or **both**.

Statement	Longitudinal waves only	Transverse waves only	Both longitudinal and transverse waves
The wave carries energy.	☐	☐	☐
The wave travels at 90° to the vibration that produced it.	☐	☐	☐

[2 marks]

Frequency and period

- The frequency of a wave is the number of waves passing a point each second.
- Frequency is measured in hertz (Hz). 1 Hz = 1 wave per second.
- The time taken for one complete wave is called its period. This is measured in seconds.
- The equation that links frequency and period is:

$$\text{period} = \frac{1}{\text{frequency}}$$

- To use the equation, frequency must be in Hz and period in seconds.

1. What is the unit for frequency?

Tick **one** box.

☐ m   ☐ m/s   ☐ Hz   [1 mark]

2. Give the name for the number of waves passing a point in one second.

_____   [1 mark]

3. Some water waves have a frequency of 10 Hz.

b Give the number of these waves that would pass a fixed point in 1 s.

_____   [1 mark]

b Give the number of these waves that would pass a fixed point in 6 s.

_____   [1 mark]

4. Calculate the period of a wave that has a frequency of 5 Hz.

$$\text{Period} = \frac{1}{\text{frequency}}$$

Support
Start by writing down the equation you are going to use. You need to know this equation.

_____

Period = _____ s   [3 marks]

5. Calculate the frequency of a wave that has a period of 0.02 s.

_____

_____

Frequency = _____ Hz   [3 marks]

Wave calculations

- The speed of a wave tells us how far the wave travels in one second. The unit is m/s.
- All waves obey the wave equation:
- wave speed = frequency × wavelength
- To use the equation, wave speed must be in m/s, frequency must be in Hz and wavelength in m.

1. A wave has a wavelength of 3 m and a frequency of 12 Hz.

Calculate the wave speed. Give the unit.

Support

You need to know this equation – it may not be given in an exam.

Remember to show all your working out.

 Wave speed = frequency × wavelength

= 12 Hz × 3 m

Wave speed = _____ Unit _____ [4 marks]

2. Calculate the speed of a wave which has a frequency of 2×10^3 Hz and a wavelength of 4×10^2 m.

Support

To multiply two numbers in standard form, multiply the whole numbers first (2 × 4). Next, multiply together the two powers of ten ($10^3 \times 10^2$). Remember, when you multiply powers of ten you **add** the indices.

_____

_____

Wave speed = _____ m/s [2 marks]

3. Calculate the speed of a wave which has a frequency of 0.5 MHz and a wavelength of 3 mm.

Support

To get the wave speed in metres per second, you must convert the wavelength to metres and the frequency to Hz.

M means mega which is 10^6. 1 mm = 10^{-3} m or 0.001 m.

 0.5 MHz = _____ Hz

3 mm = _____ m

_____

Wave speed = _____ m/s [4 marks]

4. The speed of sound in air is 340 m/s. A musical note has a wavelength of 20 cm.

Calculate the frequency of this note.

 20 cm = _____ m

Wave speed = 340 m/s = frequency in Hz × _____ m

Frequency = ────────────────

Frequency = _____ Hz [4 marks]

5. The speed of sound in water is about 1400 m/s. Calculate the wavelength of a water wave that has a frequency of 120 Hz.

_____

_____

Wavelength = _____ m   [3 marks]

6. A student did an experiment to find the speed of sound waves in air.

She placed a sound generator and a microphone 2.8 m apart. She used a datalogger to measure and record the time taken for sound to travel this distance.

Support

Remember, you need to convert the units in milliseconds (ms) to seconds (s). There are 1000 ms in 1 s.

a Write down the equation that links speed, distance travelled and time.

_____   [1 mark]

b The datalogger recorded a time of 8 ms.

Calculate the speed of sound from the student's results.

_____

_____

_____   [4 marks]

7. A student is investigating water waves in a ripple tank.

He set up the equipment shown.

a Describe how the student should use the apparatus to find the frequency of water waves in the tank.

Show Me   Count the _____

_____   [2 marks]

shallow tank of water

oscillating paddle

b The student decided to find the wavelength of the water waves by measuring the distance from one wave crest to the next.

Why should the student instead measure the distance across as many complete waves as possible, then divide the total distance by the number of complete waves?

Tick **one** box.

wave patterns on a viewing screen

[ ] To make the experiment more accurate.

[ ] To make the measurements more repeatable.

[ ] To make the experiment a fair test.   [1 mark]

c State how to use the student's measurements from parts (a) and (b) to calculate the speed of the water waves.

_____   [1 mark]

The electromagnetic spectrum

- All electromagnetic waves are transverse waves that transfer energy.
- The electromagnetic spectrum of electromagnetic waves has a continuous range wavelength and frequency.
- All types of electromagnetic wave travel at the same speed through a vacuum (space) or air.
- The main groups of waves in the electromagnetic spectrum are (in order of increasing frequency) radio, microwave, infrared, visible light (red to violet), ultraviolet, X-rays and gamma rays.
- Our eyes detect only visible light, which is a very small part of the electromagnetic spectrum.

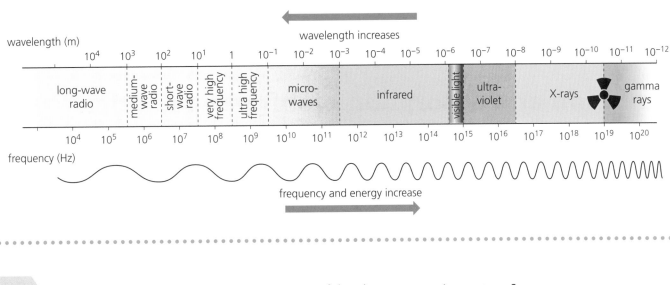

....

1. Which one of the following waves is **not** part of the electromagnetic spectrum?

Tick **one** box.

☐ microwaves ☐ sound waves ☐ ultraviolet rays

[1 mark]

2. Which of the following electromagnetic waves has the **longest** wavelength?

Tick **one** box.

☐ gamma rays ☐ radio waves

☐ light waves ☐ X-rays

[1 mark]

3. A student is asked to put some types of electromagnetic wave in order, starting with the **highest** frequency.

Her answer is:

gamma rays   X-rays   infrared   light   radio waves

a Which **two** types of electromagnetic waves has she put in the wrong order?

i. _____

ii. _____

[2 marks]

b Which **two** types of electromagnetic waves in the list transfer the most energy?

i. _____

ii. _____

[2 marks]

4. Draw **one** line from each type of electromagnetic radiation to the equipment that can be used to detect it.

Infrared radiation	Antenna
Gamma radiation	Eyes
Visible light	Geiger-Müller tube
Radio waves	Thermometer

[4 marks]

Emission and absorption of infrared radiation

- The colour and type of a surface affect how well the surface absorbs or reflects infrared radiation.
- Shiny surfaces are poor absorbers because they reflect most of the infrared radiation.

1. Explain how shiny foil blankets keep marathon runners warm after a race.

_____

_____ [1 mark]

2. A student did an experiment to find out whether a dull black surface emitted more infrared radiation than a shiny white surface.

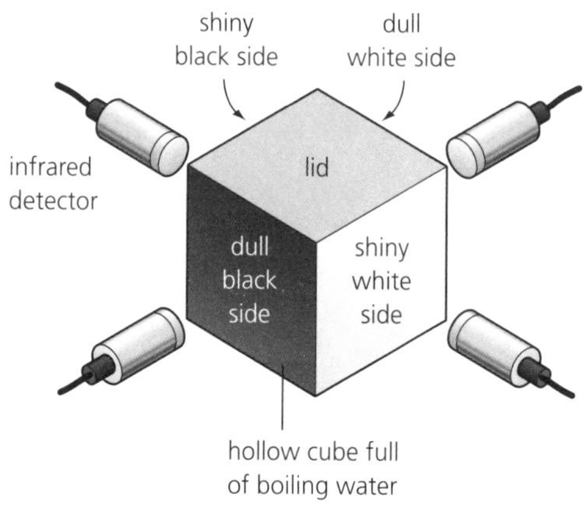

This is the method used.

i. Fill the cube with very hot water and replace the lid.

ii. Use an infrared detector to measure the amount of infrared radiated from each surface in a period of time.

shiny black side — dull white side — lid — infrared detector — dull black side — shiny white side — hollow cube full of boiling water

a Name the **dependent** variable in this investigation.

_____ [1 mark]

b Name **one** variable that should be controlled in this investigation.

_____ [1 mark]

3. Another student used different equipment for the experiment in question 2.

She used two identical containers, one painted white and one painted black. She also had very hot water and a thermometer.

Describe a method the student could use to investigate whether a black surface emitted more infrared radiation than a white surface.

Include in your description the measurements she should make, and how she can make it a fair test.

_____

_____

_____

_____ [4 marks]

Refraction

- When light travels from air into water or glass (or the other way) it can change direction. This is refraction.
- There is no refraction if the light wave crosses the boundary at 90°.
- A line in a ray diagram that is drawn at 90° (right angles) to the surface is called the normal line.
- If the ray of light enters water or glass from air it will bend towards the normal.
- If the ray of light enters air from water or glass it will bend away from the normal.

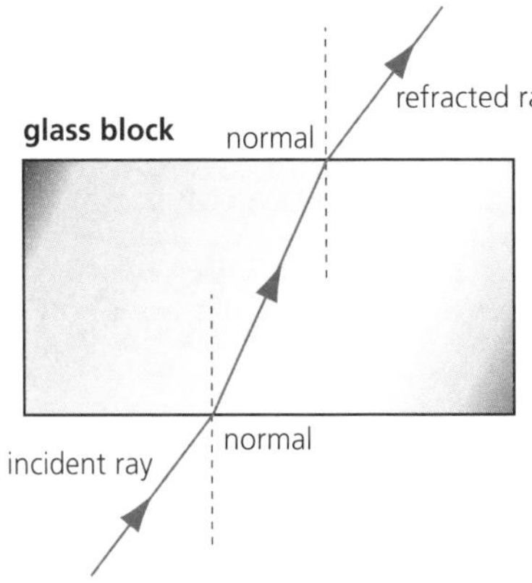

glass block   normal   refracted ray

incident ray   normal

1. Choose **one** of these phrases to complete each sentence below:

bends away from the normal	bends towards the normal	does not change direction

a When light enters a glass block from air at an angle of 45°,

the light ray _____. [1 mark]

b Light that enters a glass block along a normal _____. [1 mark]

2. The diagram shows a light ray in air reaching a glass block.

Complete the diagram to show what happens to the ray of light as it enters, travels through and leaves the glass block.

You will need to draw the normals.

[4 marks]

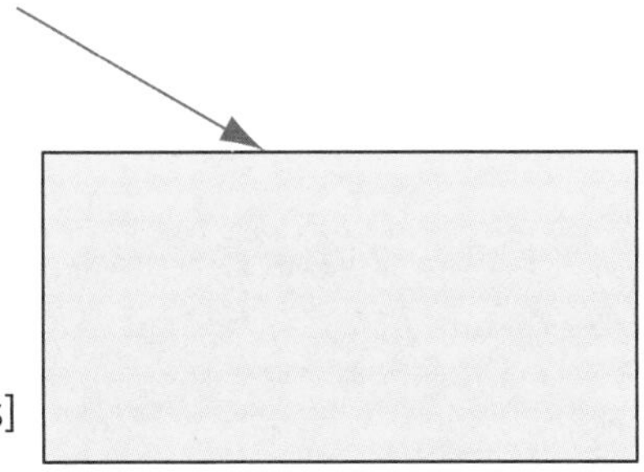

Uses and hazards of the electromagnetic spectrum

- Different types of wave in the electromagnetic spectrum have different uses.
- Different types of wave in the electromagnetic spectrum have different effects.
- The effects of electromagnetic waves on the body depend on the type and amount of radiation received.
- Radiation dose is a measure of the risk of harm resulting from an exposure of the body to the radiation.
- Ultraviolet waves, X-rays and gamma rays are the most hazardous.

1. Complete the table. Give **one** use of each type of electromagnetic radiation.

Type of electromagnetic radiation	Use
Microwaves	
Ultraviolet rays	
Gamma rays	
Infrared radiation	
Radio waves	

[5 marks]

2. Draw **one** line from each type of electromagnetic radiation below to the harmful effect it can have.

Gamma radiation		Premature skin ageing and skin cancer
Infrared radiation		Genetic damage and cancer
Ultraviolet radiation		Skin and tissue burns

[3 marks]

3. Give **two** factors that affect how much harm electromagnetic radiation can cause.

i. _____

ii. _____ [2 marks]

Magnets and magnetic forces

- Magnets attract magnetic materials. Iron, steel, cobalt and nickel are magnetic materials.
- A magnet has two poles. Magnetic forces are strongest at the poles of a magnet.
- Magnets attract and repel each other. These are non-contact forces.

two like poles repel each other

two unlike poles attract each other

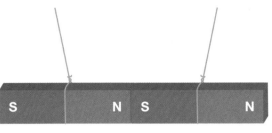

- When unmagnetised iron is near a magnet it becomes an induced magnet.
- An induced magnet loses its magnetism when moved away from the magnet – the magnetism is not permanent.

1. Which **two** of these metals are magnetic?

   Tick **two** boxes.

   ☐ aluminium   ☐ cobalt   ☐ copper   ☐ iron   ☐ tin   [2 marks]

2. Complete the sentences to describe the force between poles of two magnets.

   Two south poles will _____ each other.

   A south and a north pole will _____ each other.   [2 marks]

3. Where are the magnetic forces close to a magnet strongest?

   Tick **one** box.

   ☐ At the north pole

   ☐ At the south pole

   ☐ In between the poles

   ☐ At both poles   [1 mark]

4. **a** State the name for an object that has a magnetic field all the time.

   _____   [1 mark]

   b State the name for an object that becomes magnetised when it is near a magnet.

   _____   [1 mark]

Magnetic fields

- Magnets have a magnetic field around them.
- The shape of a magnetic field can be found using a compass or iron filings.
- The direction of the field is the direction of the force on a north pole placed at that point.
- The strength of the magnetic field depends on the distance from the magnet.
- The Earth has a magnetic field.
- A compass contains a magnet. Its arrow is a north pole. A compass can show the direction of a magnetic field.

1. **a** Draw field lines to show the shape of the magnetic field around the bar magnet below.

Include arrows to show the direction of the field. [2 marks]

> **Support**
> Note you have been asked to do **two** things for **two** marks.

b Name **two** things you could use to find the shape of the magnetic field around a bar magnet.

i. _____

ii. _____ [2 marks]

2. A question in a test asked:

'How does the way a magnetic compass works give evidence that the Earth has a magnetic field?'
A student wrote the answer below. She has made two errors.

A magnetic compass contains a small piece of steel but this is not a magnet. The Earth has a magnetic field. The compass needle moves to follow the direction of the Earth's magnetic field. The fact that the Earth has a magnetic field means its core must act like a magnet. The Earth's magnetic field doesn't have poles.

a Underline the **two** errors in the answer above. [2 marks]

b Write a correct version of each wrong sentence.

> Show Me

A magnetic compass contains _____

_____

The Earth's magnetic field has _____ [2 marks]

3. Explain why a compass needle always points in the north–south direction.

> **Support**
> In long descriptive answers, make the steps clear. You could use bullet points.

_____

_____

_____ [3 marks]

The magnetic effect of a current

- When a current flows through a wire, a magnetic field is produced around the wire.
- Coiling the wire into a solenoid increases the magnetic effect.
- The magnetic field of a solenoid looks acts like the magnetic field of a bar magnet.
- Which end is the north pole and which the south depends on the direction of the current.
- The magnetic field inside a solenoid is strong and the same strength everywhere.
- The magnetic field of a solenoid can be made stronger by increasing the current or by adding an iron core.
- An electromagnet is a solenoid with an iron core.

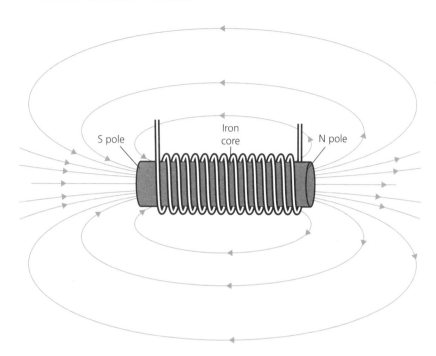

1. **a** State whether an electromagnet is a **permanent** magnet or an **induced** magnet.

   _____ [1 mark]

   b Give a reason for your answer to part (a).

   _____ [1 mark]

2. When a current flows through a solenoid, a magnetic field is produced.

   Complete the table to show how each change will affect the strength of the magnetic field of a solenoid.

   Tick **one** box in each row.

Change	Strength increases	Strength decreases	No effect
Decrease the current through the wire	☐	☐	☐
Add a copper core	☐	☐	☐
Add an iron core	☐	☐	☐

[3 marks]

3. A student is telling his friend about electromagnets. He has made three mistakes in his description.

Underline **three** mistakes.

You can make an electromagnet from a coil of wire wrapped around a piece of metal, like a large nail. You have to attach the coil of wire to a battery to pass a current through it. To make the electromagnet really strong you have to leave the metal inside the coil of wire. It doesn't matter what metal it is. Adding more turns to the solenoid will make it stronger too. Changing the size of the current won't make any difference to how strong the electromagnet is. Once you have made the electromagnet work it will keep on working even when you've switched the current off.

[3 marks]

4. A student wants to investigate the magnetic effect produced by a straight wire carrying a current.

a Describe an experiment the student could do with the equipment in the diagram, to show that a current in the straight wire has a magnetic effect.

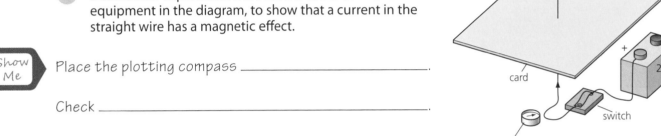

Place the plotting compass _____ .

Check _____ .

Switch on _____ and

_____ .

[3 marks]

b Describe an experiment the student could do to find the shape and the direction of the magnetic field.

_____

_____

_____

[3 marks]

5. **a** Look at the diagram below. Draw field lines to show the shape of the magnetic field inside and around the solenoid. [2 marks]

b Add arrows to the lines you have drawn to show the direction of the magnetic field inside and outside the solenoid. [2 marks]

Energy stores

1.

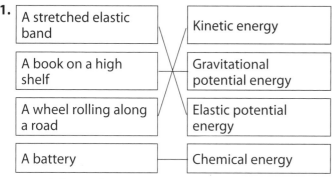

A stretched elastic band	Kinetic energy
A book on a high shelf	Gravitational potential energy
A wheel rolling along a road	Elastic potential energy
A battery	Chemical energy

1 mark for each correct line

2. The water stores energy as thermal energy. [1 mark]

Before the book falls it has a store of gravitational potential energy. [1 mark]

As the book falls, the kinetic energy increases. [1 mark]

3. The force transfers energy from a store of chemical energy [1 mark] to a store of kinetic energy. [1 mark]

As the car slows down the amount of energy in the kinetic store decreases. [1 mark]

4. When the ball is moving it stores kinetic energy. [1 mark]

The store of kinetic energy decreases. [1 mark]

The store of gravitational potential energy increases. [1 mark]

Calculating energy changes

1. Energy gained = 20 kg × 10 N/kg × 2 m [1 mark] = 400 J [1 mark]

2. Gravitational potential energy gained = 8 kg × 10 N/kg × 0.5 m [1 mark] = 40 [1 mark] J [1 mark]

3. Kinetic energy = 0.5 × 12 kg × (2 m/s)^2 [1 mark] = 24 J [1 mark]

4. Kinetic energy = 0.5 × 0.4 kg × (10 m/s × 10 m/s) [1 mark] = 20 J [1 mark]

5. a) Kinetic energy = 0.5 × mass × (speed)^2 [1 mark]

b) Kinetic energy = 0.5 × 2 × 10^3 kg × (20 m/s)^2 [1 mark] = 400 000 [1 mark] J [1 mark] (or 400 [1 mark] kJ [1 mark])

c) Four times bigger [1 mark]

6. a) Stored elastic potential energy = 0.5 × 6 N/m × (0.4 m)^2 [1 mark] = 0.48 J [1 mark]

b) Stored elastic potential energy = 0.5 × 6 N/m × (0.6 m)^2 [1 mark] = 1.08 J [1 mark]

Calculating energy changes when a system is heated

1. a) Change in thermal energy = 2 kg × 380 J/kg/°C × 5 °C [1 mark] = 3800 J [1 mark]

b) It would need more energy to heat the copper. [1 mark]

2. Change in thermal energy = 3 kg × 4200 J/kg/°C × 30 °C [1 mark] = 378 000 [1 mark] J [1 mark] (or 378 [1 mark] kJ [1 mark])

3. a) Thermal energy [1 mark]

b) 30 kJ [1 mark]

c) Change in thermal energy = 30 000 J = 1 kg × specific heat capacity × 40 °C [1 mark]

Specific heat capacity = $\dfrac{30\,000 \text{ J}}{(1\,\text{kg} \times 40\,°C)}$ [1 mark]

= 750 J/kg/°C [1 mark]

d) Insulation (around and on top of the block). [1 mark]

Calculating work done

1. Work done = force × distance moved in direction of force. [1 mark]

2. Work done = 40 N × 1.5 m [1 mark] = 60 J [1 mark]

3. 32 J [1 mark]

4. Work done = 600 N × 3 m [1 mark] = 1800 [1 mark] J or Nm [1 mark]

5. a) Work done = force used × distance travelled [1 mark]

b) 2500 J = force × 50 m [1 mark]

Force = $\dfrac{2500 \text{ J}}{50 \text{ m}}$ [1 mark] = 50 N [1 mark]

Power

1. a) Power = $\dfrac{\text{work done}}{\text{time}}$ [1 mark]

b) Power = $\dfrac{800 \text{ J}}{4 \text{ s}}$ [1 mark] = 200 W [1 mark]

2. a) Power = $\dfrac{\text{energy transferred}}{\text{time}}$ [1 mark]

b) Power = $\dfrac{600 \text{ J}}{10 \text{ s}}$ [1 mark] = 60 W [1 mark]

3. Power = $\dfrac{60\,000 \text{ J}}{20 \text{ s}}$ [1 mark] = 3000 W [1 mark] = 3 kW [1 mark]

4. a) Power = $\dfrac{200 \text{ J}}{4 \text{ s}}$ [1 mark] = 50 W [1 mark]

b) The man is working at a lower power than the boy. [1 mark]

c) The man takes a longer time to do the same amount of work [1 mark]. This means the man does less work in one second [1 mark] (and so his power is lower than the boy's power).

5. a) 2 000 000 W (or 2 × 10^6 W) [1 mark]

b) 2 MW means 2 × 10^6 joules every second. So in 1 minute (60 seconds) the energy transferred = 60 × 2 × 10^6 J [1 mark] = 120 000 000 J (or 1.2 × 10^8 J) [1 mark].

Conservation of energy

1. Energy can be destroyed. [1 mark]
2. A vacuum cleaner is designed to transfer energy from the mains electrical supply to kinetic [1 mark] energy.

   When a vacuum cleaner is used some of the input energy is wasted by transfer of thermal [1 mark] energy to the surroundings.
3. The rest of the energy (40 J) is dissipated (wasted) [1 mark] through thermal energy transfer / heating [1 mark] to the surroundings [1 mark].
4. a) Energy transferred by current = 2000 J – 800 J = 1200 J [1 mark]

   b) One of: thermal energy or sound [1 mark]

Ways of reducing unwanted energy transfers

1. Lubrication reduces unwanted energy transfers. [1 mark]
2.

Suggested change	Increase thermal transfer	Reduce thermal transfer
Make the outside walls thicker		✓ [1 mark]
Remove loft insulation	✓ [1 mark]	
Replace carpets with stone floors which are better heat conductors	✓ [1 mark]	
Add double glazing to the windows		✓ [1 mark]

3. Electric motors are designed to transfer energy from the electrical supply to kinetic [1 mark] energy.

   Not all the input energy is usefully transferred. There is friction [1 mark] between moving parts in the motor.

   This means that the motor transfers some energy to the surroundings as thermal [1 mark] energy.

   The motor can be lubricated [1 mark] to reduce the amount of unwanted energy transfer.

Efficiency

1. a) $40\% = \frac{40}{100} = 0.4$ [1 mark]

   b) $0.65 = \frac{65}{100} = 65\%$ [1 mark]

2. Efficiency $= \frac{60\,J}{100\,J}$ [1 mark] $= 60\%$ [1 mark]

3. Efficiency $= \frac{30\,W}{100\,W}$ [1 mark] $= 0.3$ [1 mark]

4. Efficiency $= \frac{6000\,J}{8000\,J}$ [1 mark] $= 75\%$ [1 mark]

Renewable and non-renewable energy resources

1.

Resource	Renewable	Non-renewable
Bio-fuel	✓ [1 mark]	
Tides	✓ [1 mark]	
Coal		✓ [1 mark]
Wind	✓ [1 mark]	

2. Renewable energy resources can be replaced. [1 mark]

3.

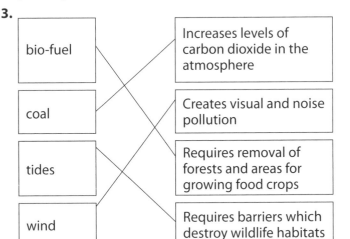

[1 mark for each correct line]

4. Any two [1 mark each] from:

   Wind turbines cannot provide electricity when the wind does not blow.

   Solar panels cannot provide electricity when the sun is not shining.

   Wave generators cannot provide electricity when the sea is calm.

5. a) Decreased (slightly) [1 mark]

   b) Increased (more than doubled) [1 mark]

   c) Increased [1 mark]

6. a) $\frac{150}{200} = \frac{1}{4}$ [1 mark]

   b) (50:150 =) 1:3 [1 mark]

   c) One of:

   Bio-fuels are renewable (or will not run out).

   Bio-fuels give less pollution. [1 mark]

   d) One of:

   As a heating fuel.

   To generate electricity. [1 mark]

Answers

Section 2: Electricity

Circuit diagrams

1. A: Shows a battery, which is two or more cells used together. [1 mark]

   B: Shows a cell. [1 mark]

2.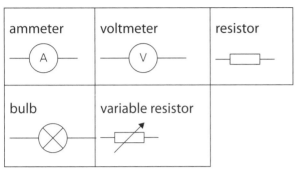

   [1 mark each]

3.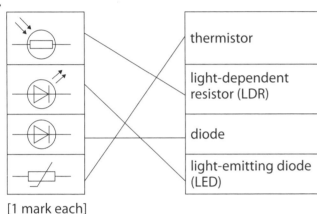

   [1 mark each]

4.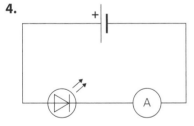

   Circuit diagram with correct symbols for cell, ammeter and LED [1 mark] joined in series (on after the other in any order) [1 mark] and with the LED the correct way round, as shown here [1 mark].

Electrical charge and current

1. Cell (or battery or power pack or power supply) [1 mark]

2. a) A2: 2 A

      A3: 2 A [1 mark for both correct]

   b) The current is the same everywhere in a series circuit. [1 mark]

3. Charge flow: coulomb, C [1 mark]

   Current: amp, A [1 mark]

   Time: second, s [1 mark]

4. a) Charge flow = current × time = 3 A × 20 s [1 mark] = 60 C [1 mark]

   b) 10 minutes × 60 = 600 seconds [1 mark]

      Charge flow = 3 A × 600 s [1 mark] = 1800 C [1 mark]

5. 25 000 C = 5 A × time [1 mark]

   $$\text{Time} = \frac{25\,000\ \text{C}}{5\ \text{A}}\ \text{[1 mark]} = 5000\ \text{s}\ \text{[1 mark]}$$

6. $5\ \text{mA} = \frac{5}{1000} = 0.005\ \text{A}$ [1 mark]

   20 minutes = 20 × 60 s = 1200 s [1 mark]

   Charge flow = 0.005 A × 1200 s [1 mark] = 6 C [1 mark]

Electrical resistance

1. The potential difference is sometimes called the voltage. [1 mark]

   If the potential difference in a circuit is increased, the current increases. [1 mark]

   If the resistance is increased for the same potential difference, the current decreases. [1 mark]

2. Potential difference = current × resistance [1 mark] = 4 A × 20 Ω [1 mark] = 80 V [1 mark]

3. 12 V = 3 A × resistance [1 mark]

   $$\text{Resistance} = \frac{12\ \text{V}}{3\ \text{A}} = 4\ \text{ohm (or } \Omega)\ \text{[1 mark]}$$

4. The ammeter and voltmeter are in the wrong places [1 mark]. The ammeter must be in series with the component [1 mark] and the voltmeter must be in parallel with the component. [1 mark]

5. a) The length of the wire. [1 mark]

   b) The resistance of the wire (or the current in the wire). [1 mark]

   c) One of: same type of wire (or wire from same reel), same temperature, same voltage. [1 mark]

6. Any three points from the following [1 mark each]:

   Measure the current and voltage for a fixed length of wire.

   Calculate resistance of wire from readings.

   Repeat for different lengths of wire.

   Use at least 5 different lengths of wire.

7. Resistance is proportional to length. [1 mark]

Changing resistances

1. The two LEDs are connected in opposite directions [1 mark]. A diode will only let current flow in one direction [1 mark], so one of them is stopping the current flowing [1 mark].

Answers

2. a)

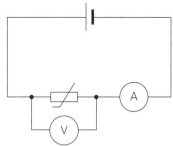

Correctly drawn circuit symbols [1 mark] with cell, ammeter and thermistor in series [1 mark] and a voltmeter connected across the thermistor [1 mark].

b) So they can calculate the resistance. [1 mark]

3. a) When the potential difference across a filament lamp increases, its resistance increases, OR: When the potential difference across a filament lamp decreases, its resistance decreases. [1 mark]

b) When the temperature of a thermistor increases, its resistance decreases. [1 mark]

c) When the brightness decreases, the resistance of an LDR increases. [1 mark]

4. [4] If there is a fire the temperature rises and so the resistance of the thermistor goes down.

[2] When the thermistor is cold its resistance is high.

[1] The circuit includes a cell, a thermistor and a buzzer.

[5] This allows a bigger current to flow in the circuit, so the buzzer sounds.

[3] Only a very small current can flow in the circuit, so the buzzer does not sound. [1 mark each]

5. When the temperature is low, the resistance of the thermistor is high [1 mark] so no (or only a small) current flows and the cooling system is not switched on [1 mark].

If the temperature rises, the resistance of the thermistor goes down [1 mark] so a current can power the cooling system (or similar words) [1 mark].

6. A thermistor has a low resistance when the temperature is high, so the current is also high [1 mark]. We want the heater on when the temperature is low – when the thermistor's resistance is high [1 mark].

7. The resistance of the LDR becomes high when it gets dark [1 mark], so only a small current flows [1 mark]. The lights are switched on by a special switch that works when the current is low [1 mark].

Series and parallel circuits

1. a) A, D [1 mark]

b) B, C [1 mark]

2. a) 4 V [1 mark]

b) In a series circuit the potential difference is divided between the components. [1 mark]

3. 6 V [1 mark]

4. a) 1.5 A [1 mark]

b) In a series circuit the current is the same in all components. [1 mark]

5. a) 12 V [1 mark]

b) In a parallel circuit the potential difference across each branch is the same (as the potential difference across the cell). [1 mark]

6. a) 2 A [1 mark]

b) The current though the cell is the sum of the currents through the separate branches. [1 mark]

7. 12 Ω [1 mark]

Mains electricity

1. a) A direct potential difference always acts in the same direction [1 mark]. With an alternating potential difference the direction changes many times a second. [1 mark]

b) Cell or battery or power pack. [1 mark]

2. A is the neutral wire. [1 mark]
B is the earth wire. [1 mark]
C is the live wire. [1 mark]

3. a) It will carry a current. [1 mark]

b) The person would provide a connection between the live supply and earth [1 mark]. The live wire is at a potential difference of 230 V compared to earth, [1 mark] so a current could flow through the person [1 mark] and they would get burns and/or an electric shock [1 mark].

Energy changes in circuits

1. Energy transferred every second = 2000 J [1 mark]

2. The energy transferred by an appliance is greater if the appliance has a higher power [1 mark] or if it is switched on for a longer time. [1 mark]

The work done (energy transferred) by an electric current is greater [1 mark] if more charge flows or if the charge is 'pushed' by a higher potential difference. [1 mark]

3. Charge: coulomb, C [1 mark]
Power: watt, W [1 mark]
Energy: joule, J [1 mark]
Time: second, s [1 mark]

4. Energy transferred = charge flow × potential difference [1 mark] = 1 C × 230 V [1 mark] = 230 J [1 mark]

5. a) Energy transferred = power × time [1 mark]

b) 5 minutes = 5 × 60 s = 300 s [1 mark]
Energy transferred = 3000 W × 300 s [1 mark] = 900 000 J [1 mark]

c) 900 kJ [1 mark]

6. 50 W lamp: Energy transferred = power × time [1 mark] = 50 W × 500 s [1 mark] = 25 000 J [1 mark]

40 W lamp: Energy transferred = 40 W × 625 s [1 mark] = 25 000 J [1 mark]

7. 500 J = charge flow × 230 V [1 mark]

Charge flow = $\frac{500\ J}{230\ V}$ [1 mark] = 2.17 (or 2.2) C [1 mark]

8. a) Kettle A [1 mark] transfers the most energy each second because it has a higher power rating [1 mark].

b) Both kettles transfer the same amount of energy to boil the same volume of water [1 mark]. Although kettle B has a lower power rating, it is switched on for longer [1 mark].

c) 3 minutes = 3 × 60 s = 180 s [1 mark]

Energy transferred = 2000 W × 180 s [1 mark] = 360 000 J [1 mark]

d) Energy transferred = 360 000 J = charge flow × 230 V [1 mark]

Charge flow = $\frac{360\ 000\ J}{230\ V}$ [1 mark]

= 1565 (or 1600) C [1 mark]

Electrical power

1. Power: watt, W [1 mark]

Current: amp, A [1 mark]

Potential difference: volt, V [1 mark]

Resistance: ohm, Ω [1 mark]

2. Power = potential difference × current [1 mark] = 12 V × 4 A [1 mark] = 48 W [1 mark]

3. Power = 20 W = 230 V × current [1 mark]

Current = $\frac{20\ W}{230\ V}$ [1 mark] = 0.087 (or 0.09) A

[1 mark]

4. Power = (current)² × resistance [1 mark] = (10 A)² × 25 Ω [1 mark] = 2500 W [1 mark]

5. a) Power = 40 W = potential difference × 0.4 A [1 mark]

Potential difference = $\frac{40\ W}{0.4}$ [1 mark] = 100 V [1 mark]

b) Power = 40 W = (0.4 A)² × resistance [1 mark]

Resistance = 40 W/(0.4 A)² [1 mark] = 250 Ω [1 mark]

The national grid

1. a) step-up [1 mark]

b) step-down [1 mark]

c) step-down [1 mark]

2.

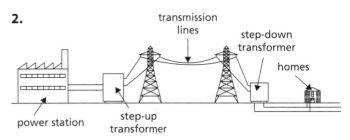

[1 mark for each]

3. a) Less energy is wasted by heating. [1 mark]

b) It is safer. [1 mark]

4. a) 33 000 V [1 mark]

b) 400 000 V [1 mark]

Section 3: Particle model of matter

Matter and density

1. Mass: kilograms (or kg) [1 mark]

Volume: cubic metres (or m³) [1 mark]

Density: kilograms per metre cubed [1 mark] (or kg/m³)

2. A: liquid [1 mark]

B: solid [1 mark], most dense [1 mark] (either order)

C: gas [1 mark], least dense [1 mark] (either order)

3. The particles are closest together when it is solid [1 mark], so there is more mass [1 mark] in a certain volume [1 mark].

4. Density = $\frac{mass}{volume}$ [1 mark] = $\frac{183\ 000\ kg}{200\ m^3}$ [1 mark]

= 915 kg/m³ [1 mark]

5. Density = 2000 kg/m³ = $\frac{mass\ in\ kg}{0.002\ m^3}$ [1 mark]

Mass = 2000 kg/m³ × 0.002 m³ [1 mark] = 4 kg [1 mark]

6. Fill the displacement can with water until it comes up to the spout [1 mark]. Put the object in the water and catch the water that spills in a measuring cylinder [1 mark]. The volume of water is the volume of the object [1 mark].

Changes of state

1. a) solid → gas [1 mark]

b) liquid → gas [1 mark]

c) gas → liquid [1 mark]

2. a) melting [1 mark]

b) freezing [1 mark]

3. The number stays the same. [1 mark]

4. a) physical change [1 mark]

b) chemical change [1 mark]

c) both boxes ticked [1 mark]

d) physical change [1 mark]

5. internal energy [1 mark]

6. a)

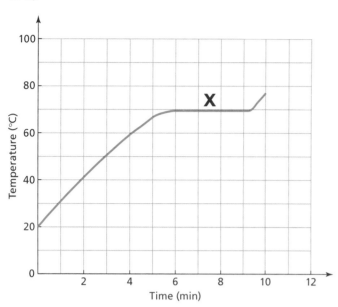

[1 mark]

b) 70 °C [1 mark]

c)

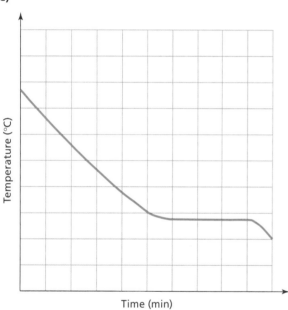

Line starts at a high temperature and descends to the right [1 mark] with a horizontal section in the middle [1 mark].

Latent heat

1. a) It boils. [1 mark]

b) The energy transferred to the water is increasing the internal energy [1 mark], which is changing the water from a liquid to a gas [1 mark].

2. a) liquid to gas [1 mark]

b) solid to liquid [1 mark]

3. Mass in kg = 500 g/1000 = 0.5 kg [1 mark]

Energy needed = 0.5 kg × 334 kJ/kg [1 mark] = 167 kJ [1 mark]

4. At first the temperature of the water will go down. [1 mark]

This is because it is warmer than the temperature in the freezer. [1 mark]

While ice is forming the temperature will stay the same (or stay at 0 °C) until all the water has frozen. [1 mark]

Then the temperature will go down again (until it is the same temperature as the inside of the freezer). [1 mark]

Gas pressure

1. a) correct [1 mark]

b) correct [1 mark]

c) correct [1 mark]

d) incorrect [1 mark]

e) incorrect [1 mark]

2. When the temperature goes down the kinetic energy of the particles decreases. OR When the temperature goes up the kinetic energy of the particles increases. [1 mark]

The force on the container walls is greater if the particles are moving more quickly. OR The force on the container walls is smaller if the particles are moving more slowly. [1 mark]

3. When the temperature goes down the particles have less internal energy [1 mark] so they are moving more slowly [1 mark]. They hit the walls of the container less often [1 mark] and with less force [1 mark] (so the total force from the particles is less).

Section 4: Atomic structure

Protons, neutrons and electrons

1. Atoms contain positively charged particles called protons. [1 mark]

Most atoms also contain uncharged particles called neutrons. [1 mark]

There are negatively charged particles called electrons [1 mark] surrounding the nucleus of an atom.

2. electron [1 mark]

3. A neutral atom has the same number of protons [1 mark] and electrons [1 mark]. (either order)

4. Atoms have equal numbers of positive and negative charges [1 mark]. The charges cancel out to make the atom neutral [1 mark].

The size of atoms

1. 0.1 nanometre [1 mark]

2. 2.6×10^{-9} m = 2.6 nm [1 mark]

3. Convert m to nm: diameter of one atom, 2×10^{-10} m = 0.2 nm [1 mark]

Number of atoms in 1 nm = $\frac{1}{0.2}$ = 5 [1 mark]

4. 10 000 times bigger [1 mark]

Elements and isotopes

1. 16 – 8 = 8 [1 mark]
2. a) 8 [1 mark]
   b) 18 – 8 = 10 [1 mark]
3. a) 11 [1 mark]
   b) 14 + 11 = 25 [1 mark]
   c) $^{25}_{11}$Na [1 mark]
4. a) 3 [1 mark]
   b) 4 [1 mark]

Electrons and ions

1. A positive ion [1 mark]
2. The atom loses one or more outer electron(s). [1 mark]
3. Number of protons = 11 [1 mark]
   Number of neutrons = 12 [1 mark]
   Number of electrons = 10 [1 mark]
4. Atoms have equal numbers of positive and negative charges so an atom is neutral. [1 mark] Electrons are negatively charged. [1 mark] Therefore, if the atom loses an electron it will have more positive charge than negative charge. [1 mark]

Discovering the structure of the atom

1. The 'pudding' part of the atom is a ball of positive [1 mark] charge.
   The 'plums' are particles with a negative [1 mark] charge.
   These particles are called electrons. [1 mark]
2. Bohr – Electrons at different energy levels [1 mark]
   Chadwick – Neutrons inside the nucleus [1 mark]
   Geiger and Marsden – The nucleus [1 mark]
3. a)

| plum pudding model [1 mark] | Positive charge spread out everywhere inside |
| nuclear model [1 mark] | Positive charge concentrated in a small space |

   b) In Bohr's model electrons surround the nucleus only in certain allowed orbits. [1 mark]

Radioactivity

1. Nuclear radiation comes from the nucleus of an unstable atom. [1 mark]

2.

Description	Name of nuclear radiation
An electron	beta particle [1 mark]
The nucleus of a helium atom (2 protons and 2 neutrons)	alpha particle [1 mark]
Electromagnetic radiation	gamma ray [1 mark]

3. A beta particle is negatively charged. [1 mark]
   A beta particle is much smaller than an alpha particle. [1 mark]
4. The count rate is the number of radioactive decays in a given time (usually in one minute or in one second). [1 mark]

Comparing alpha, beta and gamma radiation

1. Alpha particles are stopped by 5 cm of air. [1 mark]
2.

Radiation type	Distance travelled in air	Ionising power
alpha	least [1 mark]	greatest [1 mark]
beta	medium [1 mark]	medium
gamma	greatest	least [1 mark]

3. Gamma is the only type of radiation that will penetrate through the patient's body [1 mark] to reach the detector [1 mark].
4. a) First, set up the Geiger-Müller (GM) tube to measure the count rate when there is no aluminium between the source and GM tube. [1 mark]
   Then add a sheet of aluminium and measure the thickness and the new count rate. [1 mark]
   Next, change the thickness of aluminium and measure the new count rate. [1 mark]
   Continue increasing the thickness of aluminium until no radiation is detected (or the count rate stops decreasing). [1 mark]
   b) Repeat the experiment several times and find the mean of the thickness of aluminium that stops the radiation. [1 mark]

Radioactive decay equations

1. Alpha – Charge decreases by 2 [1 mark]
   Beta – Charge increases by 1 [1 mark]
   Gamma – No change [1 mark]
2. A beta particle is emitted when a neutron becomes a proton. [1 mark]
3. The mass of a nucleus decreases by 4 when an alpha particle is emitted. [1 mark]

4. $^{235}_{92}U \rightarrow \, ^{4}_{2}He \, + \, ^{231}_{90}Th$ [1 mark for each correct number]

5. $^{219}_{86}Rn \rightarrow \, ^{4}_{2}He \, + \, ^{215}_{84}Po$ [1 mark for each correct number]

6. $^{14}_{6}C \rightarrow \, ^{0}_{-1}e \, + \, ^{14}_{7}N$ [1 mark for each correct number]

Half-lives

1. The time it takes for the number of nuclei in a sample to halve. [1 mark]

2. a) Mass = 100 g [1 mark]

   b) Mass = 50 g [1 mark]

   c) Mass = 25 g [1 mark]

3. The activity will fall from 400 to 200 to 100 counts per second in 2 half-lives [1 mark]. So half-life = 1 hour [1 mark].

4. The activity halves from 80 to 40 (or 60 to 30, or 40 to 20, etc.) counts per second in 2 hours (and correct lines drawn on graph) [1 mark]. So half-life = 2 hours [1 mark].

5. a)

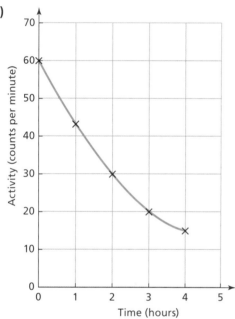

All points plotted correctly to within ± ½ small square. [2 marks]

Best-fit line should be a smooth curve. [1 mark]

   b) Activity = 24 (±1) counts per minute. [1 mark]

   c) Half-life = 2 hours (and suitable lines drawn on graph to determine this). [2 marks]

Radioactive hazards

1. The level of hazard depends on the type of radioactivity emitted. [1 mark]

2. Alpha particles are the most ionising. [1 mark]

3. a) To shield / absorb gamma radiation / prevent gamma radiation reaching workers outside the room. [1 mark]

   b) There is no radioactive material on the food. [1 mark]

4. The box protects you from alpha and beta particles [1 mark] because they cannot pass through the wood and the lead [1 mark]. The box gives some protection against gamma radiation [1 mark] but you may not be completely protected because not all the gamma rays would be stopped by the box [1 mark].

Section 5: Forces

Forces

1. velocity [1 mark]

2. mass S [1 mark], force V [1 mark], energy S [1 mark], time S [1 mark], velocity V [1 mark], momentum V [1 mark], temperature S [1 mark], acceleration V [1 mark]

3. a) electrostatic force [1 mark]

     magnetic force (or magnetism) [1 mark]

     gravitational force (or gravity) [1 mark] (any order)

   b) Two of: friction, normal contact force, water resistance, tension [1 mark each]

4. a)

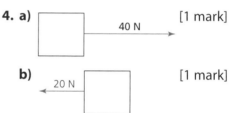

[1 mark]

   b)

[1 mark]

5. a) Resultant = 1000 N + 200 N = 1200 N [1 mark] acting backwards [1 mark]

   b) Resultant = 4000 N – 400 N = 3600 N [1 mark] acting forwards [1 mark]

   c) Resultant = 300 N – 300 N = 0 N [1 mark]

   d) Resultant = 600 N + 50 N – 100 N = 550 N [1 mark] acting backwards [1 mark]

Speed and velocity

1. a) Both, because the speed stays at 15 mph. [1 mark]

   b) Only car A, because its direction also remains constant. [1 mark]

   c) i) 50 m [1 mark]

      ii) 0 m [1 mark]

2. walking – 1.5 m/s [1 mark]

   cycling – 6 m/s [1 mark]

   running – 3 m/s [1 mark]

   speed of sound – 330 m/s [1 mark]

   car in town – 10 m/s [1 mark]

   train – 40 m/s [1 mark]

3. 5 minutes = 5 × 60 s = 300 s [1 mark]

Distance = speed × time [1 mark] = 5 m/s × 300 s [1 mark] = 1500 m [1 mark]

4. 15 minutes = 15 × 60 s = 900 s [1 mark]

Distance = 4500 m = average speed in m/s × 900 s [1 mark]

Average speed = $\frac{4500\,m}{900\,s}$ [1 mark] = 5 m/s [1 mark]

5. 10 km = 10 000 m [1 mark]

Distance = 10 000 m = 20 m/s × time [1 mark]

Time = $\frac{10\,000\,m}{20\,m/s}$ [1 mark] = 500 s [1 mark]

6. I will use a tape measure to measure a distance of (any fixed distance of several metres). [1 mark]

I will use a stopwatch to measure how long it takes a student to walk that distance. [1 mark]

I will calculate the speed by dividing the distance by the time. [1 mark]

Acceleration

1. Acceleration: m/s² [1 mark]

Change in velocity: m/s [1 mark]

Time: s [1 mark]

2. Acceleration = $\frac{change\ in\ velocity}{time\ taken}$ [1 mark]

= $\frac{10\,m/s}{5\,s}$ [1 mark] = 2 m/s² [1 mark]

3. Change in velocity = final velocity – initial velocity

= 5 m/s – 20 m/s = –15 m/s [1 mark]

Acceleration = $\frac{-15\,m/s}{5\,s}$ [1 mark] = –3 m/s² [1 mark]

4. Change in velocity = 0 – 20 m/s = –20 m/s [1 mark]

Acceleration = $-5\,m/s^2 = \frac{-20\,m/s}{time}$ [1 mark]

Time = $\frac{-20\,m/s}{-5\,m/s^2}$ [1 mark] = 4 s [1 mark]

Motion graphs

1. Distance = 30 m [1 mark]

2. a) 10 s to 15 s [1 mark]

b) 15 s to 20 s [1 mark]

3. Speed = gradient = $\frac{change\ in\ distance}{change\ in\ time}$ [1 mark]

= $\frac{(70\,m-30\,m)}{(20\,s-15\,s)} = \frac{40\,m}{5\,s}$ [1 mark]

= 8 m/s [1 mark]

4. Speed = $\frac{(90\,m-70\,m)}{(30\,s-20\,s)} = \frac{20\,m}{10\,s}$ [1 mark] = 2 m/s [1 mark]

5.

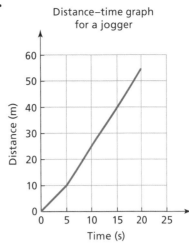

Distance–time graph for a jogger

Points correctly plotted and joined with straight lines. [2 marks (1 mark if one point mis-plotted)]

6. Velocity = 15 m/s [1 mark]

7. a) 0 to 5 s, or 60 to 65 s [1 mark]

b) 35 s to 40 s, or 80 s to 100 s [1 mark]

c) 5 s to 35 s, or 40 s to 60 s, or 65 s to 80 s [1 mark]

8. Acceleration = gradient = $\frac{change\ in\ velocity}{change\ in\ time}$ [1 mark]

= $\frac{20\,m/s}{5\,s}$ [1 mark] = 4 [1 mark] m/s² [1 mark]

9. a) Acceleration = $\frac{-25\,m/s}{20\,s}$ [1 mark] = –1.25 m/s² [1 mark]

b) The velocity is decreasing (getting less). [1 mark]

10.

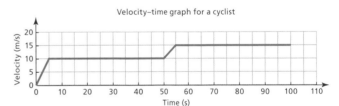

Velocity–time graph for a cyclist

Points correctly plotted and joined with straight lines. [2 marks (1 mark if one point mis-plotted)]

Acceleration and distance

1. Final velocity: m/s [1 mark], initial velocity: m/s [1 mark]

Acceleration: m/s² [1 mark]

Distance: m [1 mark]

2. Final velocity = 10 m/s, initial velocity = 5 m/s

(final velocity)² – (initial velocity)² = 2 × acceleration × distance

(10 m/s)² – (5 m/s)² = 2 × acceleration × 90 m [1 mark]

$100 - 25 = 2 \times 90 \times$ acceleration

$75 = 180 \times$ acceleration

Acceleration $= \dfrac{75}{180}$ [1 mark] $= 0.42$ [1 mark] m/s^2

3. Final velocity = 10 m/s, initial velocity = 25 m/s

$(10 \text{ m/s})^2 - (25 \text{ m/s})^2 = 2 \times -6 \text{ m/s}^2 \times$ distance [1 mark]

$100 - 625 = -12 \times$ distance

Distance $= \dfrac{-525}{-12}$ [1 mark] $= 43.75$ m [1 mark]

4. 1 km = 1000 m [1 mark]

$(40 \text{ m/s})^2 - (0 \text{ m/s})^2 = 2 \times 1000 \text{ m} \times$ acceleration [1 mark]

$1600 = 2000 \times$ acceleration

Acceleration $= \dfrac{1600}{2000}$ [1 mark] $= 0.8$ m/s^2 [1 mark]

Forces and acceleration

1. Weight = mass × gravitational field strength [1 mark]
= 15 kg × 10 N/kg [1 mark] = 150 [1 mark] N [1 mark]

2. Weight = 90 N = mass × 10 N/kg [1 mark]

Mass $= \dfrac{90 \text{ N}}{10 \text{ N/kg}}$ [1 mark] = 9 kg [1 mark]

3. The speed of the car will stay the same [1 mark] because the resultant force is zero [1 mark].

4. If she is speeding up, the resultant force on her must be forwards [1 mark] so the backwards force must be less than the forwards force [1 mark].

5. force $\propto$ acceleration [1 mark]

6. If the force doubles, the acceleration doubles. [1 mark]
If the mass is only half as big, the acceleration doubles. [1 mark]

7. a) Force = mass × acceleration [1 mark]
= 200 kg × 9 m/s^2 [1 mark] = 1800 N [1 mark]

b) Force = 1000 N = 200 kg × acceleration [1 mark]

Acceleration $= \dfrac{1000 \text{ N}}{200 \text{ kg}}$ [1 mark] = 5 m/s^2 [1 mark]

8. a) They could use light gates to measure the speed of the trolley at two places on the ramp [1 mark] and the time it takes the trolley to go between the light gates [1 mark]. Then they calculate the acceleration from the change in speed divided by time [1 mark].

b) B [1 mark]

c) Systematic error [1 mark], because the error will have the same effect on all the measurements [1 mark].

Free fall

1. a) Air resistance [1 mark]

b) It increases. [1 mark]

2. 800 N in a downwards direction [1 mark]

3. The mass of the skydiver does not change [1 mark] so the acceleration depends only on the resultant force [1 mark]. The resultant force gets smaller as she falls [1 mark], so her acceleration gets smaller [1 mark].

Action and reaction forces

1. 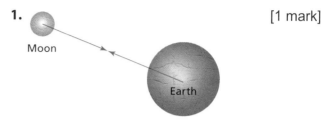 [1 mark]

2. a)

b)

[1 mark for each correct arrow drawn]

3. 400 N [1 mark]

4. a) balanced forces [1 mark]

b) action and reaction pair [1 mark]

c) balanced forces [1 mark]

d) action and reaction pair [1 mark]

Forces and braking

1. When a vehicle brakes, friction [1 mark] causes the kinetic [1 mark] energy of the vehicle to be reduced (or made smaller) [1 mark].

The greater the speed of the vehicle, the greater [1 mark] the braking force needed to stop it in a certain distance [1 mark].

2. Energy transferred by braking may make the brakes overheat [1 mark]. Large forces may make the driver lose control [1 mark].

3. a) Speed $\approx$ 10 m/s [1 mark]

b) i) Acceleration $= \dfrac{\text{change in velocity}}{\text{time taken}}$ [1 mark]

ii) Acceleration $= \dfrac{10 \text{ m/s}}{5 \text{ s}}$ [1 mark] = 2 m/s^2 [1 mark]

c) i) Force = mass × acceleration [1 mark]

ii) Force = 1500 kg × 2 m/s^2 [1 mark] = 3000 N [1 mark]

Stopping distances

1. 0.5 s [1 mark]

2. The ruler method can be done with simple apparatus, whereas the computer method requires a computer and the correct software [1 mark].

However the computer method is more realistic because it models a real situation [1 mark].

I think the computer method is better because it gives a reaction time for responding when driving [1 mark].

3. a) 10 m [1 mark]

   b) 15 m [1 mark]

   c) 10 m + 15 m = 25 m [1 mark]

4. a) Worn brakes increase the braking distance. [1 mark]

   b) A wet road increases the braking distance. [1 mark]

   c) A driver using a mobile phone increases the thinking distance. [1 mark]

5. a) Thinking distance = reaction time × speed [1 mark]

   The reaction time is constant [1 mark], so if the speed doubles the thinking distance doubles (or, the thinking distance is proportional to the speed) [1 mark].

   b) 24 m [1 mark]

   c) Thinking distance = reaction time × speed [1 mark] = 1.2 s × 20 m/s [1 mark] = 24 m [1 mark]

6. The marks are in three bands according to the level of response:

   Level 3 [5–6 marks]: A detailed explanation, written in a sequence that makes sense.

   Level 2 [3–4 marks]: An explanation of the effect of speed on stopping distance and some relevant statements about speed limits. The statements may not be in an order that makes sense.

   Level 1 [1–2 marks]: The answer is weak. Some understanding is shown, such as simple relevant comments in no particular order.

   Points that should be made:
   - The stopping distance of a vehicle is made up of the thinking distance and the braking distance.
   - At faster speeds the thinking distance increases (in proportion to the speed).
   - The braking distance increases (with the square of the speed).
   - So at 70 mph the stopping distance is much greater than at 20 mph.
   - Outside a school there are likely to be lots of hazards (such as children crossing the road).
   - So a vehicle may need to stop quickly.
   - Therefore the speed limit is low.
   - On a motorway there are fewer hazards of this kind.
   - So it is safer for vehicles to travel at a speed that will require a longer stopping distance.

Force and extension

1. a) It will move along the surface. [1 mark]

   b) If there is only one force, it will make the object move. [1 mark]

2. E: The object goes back to its original shape when the forces are removed. [1 mark]

   I: The object does not go back to its original shape when the forces are removed. [1 mark]

3. a) The extension is how far the spring stretches [1 mark], so you need to know the stretched length and the original length to work out the extension. [1 mark]

   b) 24.9 [1 mark]

   c) Repeating measurements lets you spot any results that may be wrong [1 mark]. Calculating the mean reduces the effect of random errors [1 mark] and so you will obtain more accurate results [1 mark].

   d)

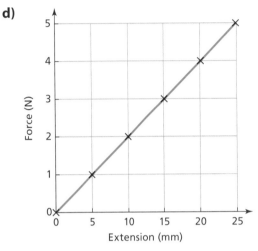

   All points plotted correctly [2 marks, if one incorrect 1 mark], line of best fit through the points [1 mark].

   e) The accuracy should improve because it will be easier to read the length/extension [1 mark] and so there will be fewer random errors [1 mark].

4. $5 \text{ cm} = \dfrac{5}{100} = 0.05 \text{ m}$ [1 mark]

   Force = spring constant × extension [1 mark]
   = 20 N/m × 0.05 m [1 mark] = 1 N [1 mark]

5. Force = 6 N = spring constant × 0.3 m [1 mark]

   Spring constant $= \dfrac{6 \text{ N}}{0.3 \text{ m}}$ [1 mark] = 20 [1 mark] N/m

   [1 mark]

6. Spring A has the largest spring constant [1 mark] because it takes a greater force to produce a certain extension [1 mark].

Springs and energy

1. The force needed to stretch the spring by 1 m. [1 mark]

2. $20 \text{ cm} = \dfrac{20}{100} = 0.2 \text{ m}$ [1 mark]

   Elastic potential energy = 0.5 × spring constant × (extension)^2
   = 0.5 × 15 N/m × (0.2 m)^2 [1 mark] = 0.3 J [1 mark]

3. Extension = 50 cm – 20 cm = 30 cm [1 mark]
   = 0.3 m [1 mark]
   Elastic potential energy = 0.5×100 N/m $\times (0.3$ m$)^2$
   [1 mark] = 4.5 J [1 mark]
4. Energy stored = 20 J = $0.5 \times$ spring constant $\times$
   $(0.6$ m$)^2$ [1 mark]
   $$\text{Spring constant} = \frac{20 \text{ J}}{0.5 \times (0.6 \text{ m})^2} \text{ [1 mark]}$$
   $\frac{20}{0.18}$ [1 mark] = 111 N/m [1 mark]

Section 6: Waves

Transverse and longitudinal waves

1. Longitudinal [1 mark]
2. Energy [1 mark]
3. Longitudinal waves have compressions and rarefactions. [1 mark]
4.

Statement	Longitudinal waves only	Transverse waves only	Both longitudinal and transverse waves
The wave carries energy.			✓
The wave travels at 90° to the vibration that produced it.		✓	

[1 mark each]

Frequency and period

1. Hz [1 mark]
2. Frequency [1 mark]
3. a) 10
   b) 60
4. Period $= \dfrac{1}{\text{frequency}}$ [1 mark]
   $= \dfrac{1}{5 \text{ Hz}}$ [1 mark]
   $= 0.2$ s [1 mark]
5. Period $= 0.02$ s $= \dfrac{1}{\text{frequency}}$ [1 mark]
   Frequency $= \dfrac{1}{0.02 \text{ s}}$ [1 mark] = 50 Hz [1 mark]

Wave calculations

1. Wave speed = frequency × wavelength [1 mark]
   = 12 Hz × 3 m [1 mark] = 36 [1 mark] m/s [1 mark]
2. Wave speed = 2×10^3 Hz $\times 4 \times 10^2$ m [1 mark]
   = $(2 \times 4) \times (10^3 \times 10^2) = 8 \times 10^5$ m/s [1 mark]
3. 0.5 MHz = 0.5×10^6 Hz [1 mark]; 3 mm = 3×10^{-3} m [1 mark]
   Wave speed = 0.5×10^6 Hz $\times 3 \times 10^{-3}$ m [1 mark]
   = 1.5×10^3 m/s [1 mark]
4. 20 cm = 0.2 m [1 mark]
   Wave speed = 340 m/s = frequency in Hz × 0.2 m [1 mark]
   $\text{Frequency} = \dfrac{340 \text{ m/s}}{0.2 \text{ m}}$ [1 mark] = 1700 Hz [1 mark]
5. Speed = 1482 m/s = 120 Hz × wavelength in m [1 mark]
   $\text{Wavelength in m} = \dfrac{1482}{120}$ [1 mark] = 12.35 m [1 mark]
6. a) Distance travelled = speed × time taken [1 mark]
   b) 8 ms = 8×10^{-3} s [1 mark]
   Distance = 2.8 m = speed $\times 8 \times 10^{-3}$ s [1 mark]
   $\text{Speed} = \dfrac{2.8 \text{ m}}{0.008 \text{ s}}$ [1 mark] = 350 m/s [1 mark]
7. a) Count the number of waves passing a point in a certain time (say 10 seconds). [1 mark]
   Divide the number of waves counted by 10. [1 mark]
   b) To make the experiment more accurate. [1 mark]
   c) Multiply together the frequency and the wavelength to get the wave speed. [1 mark]

The electromagnetic spectrum

1. Sound waves [1 mark]
2. Radio waves [1 mark]
3. a) Infrared [1 mark], light [1 mark]
   b) Gamma rays [1 mark], X-rays [1 mark]
4. Infrared radiation – thermometer [1 mark]
   Gamma radiation – Geiger-Müller tube [1 mark]
   Visible light – eyes [1 mark]
   Radio waves – antenna [1 mark]

Emission and absorption of infrared radiation

1. The shiny blanket reflects infrared radiation (thermal energy) from the runner's warm body back to the body. [1 mark]
2. a) The type of surface (dull black or shiny white). [1 mark]
   b) The distance of the detector from the surface. [1 mark]

3. Put the same amount [1 mark] of the same temperature [1 mark] hot water into each container.

   Measure the temperature of the water in each container at the start. [1 mark]

   Measure the temperature of each again after a set time. [1 mark]

Refraction

1. **a)** Bends towards the normal. [1 mark]

   b) Does not change direction. [1 mark]

2.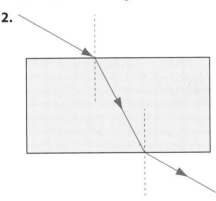

   Normals drawn [1 mark]; ray bending towards normal on entering block [1 mark]; ray bending away from normal on leaving the block [1 mark] by the same amount so emergent ray is parallel to incident ray [1 mark].

Uses and hazards of electromagnetic radiation

1.

Type of electro-magnetic radiation	Use
Microwaves	Cooking; satellite communication
Ultraviolet rays	Security marking; detecting forged bank notes; producing vitamin D in the body; killing microbes; attracting insects; getting a sun tan; fluorescence lamps
Gamma rays	Medical imaging; medical treatments; killing microbes; irradiating food
Infrared radiation	Remote control devices; thermal imaging; heating/cooking
Radio waves	TV and radio broadcasting; communications networks

   [1 mark for one use of each type]

2. Gamma radiation – genetic damage and cancer [1 mark]

   Infrared radiation – skin and tissue burns [1 mark]

Ultraviolet radiation – premature skin ageing and skin cancer [1 mark]

3. The type of radiation [1 mark] and the amount of radiation received (the dose) [1 mark].

Section 7: Magnetism and electromagnetism

Magnets and magnetic forces

1. Cobalt [1 mark], iron [1 mark]

2. Two south poles will repel [1 mark] each other.

   A south and a north pole will attract [1 mark] each other.

3. At both poles [1 mark]

4. **a)** permanent magnet [1 mark]

   b) induced (or temporary) magnet [1 mark]

Magnetic fields

1. **a)**

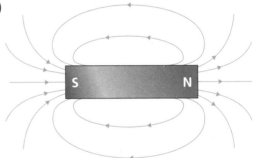

   Clear diagram showing the shape of the field lines [1 mark]; all arrows pointing in the correct direction [1 mark].

   b) (Plotting) compass [1 mark], iron filings [1 mark]

2. **a)** A magnetic compass contains a small piece of steel <u>but this is not a magnet</u> [1 mark]. The Earth has a magnetic field. The compass needle moves to follow the direction of the Earth's magnetic field. The fact that the Earth has a magnetic field means its core must act like a magnet. <u>The Earth's magnetic field doesn't have poles</u> [1 mark].

   b) A magnetic compass contains a small piece of steel that is a (permanent) magnet. [1 mark]

   The Earth's magnetic field has two poles (a magnetic north pole and a magnetic south pole). [1 mark]

3. A compass always points north–south because it is a magnet [1 mark]. The magnet has two poles [1 mark]. One of the poles (the arrow end) is attracted to / seeks the / experiences a force towards Earth's (geographic) north pole [1 mark].

The magnetic effect of a current

1. **a)** An induced magnet [1 mark]

   b) It loses its magnetism / becomes unmagnetised when current stops / is switched off. [1 mark]

2.

Change	Strength increases	Strength decreases	No effect
Decrease the current through the wire		✓	
Add a copper core			✓
Add an iron core	✓		

[1 mark each]

3. You can make an electromagnet from a coil of wire wrapped around a piece of metal, like a large nail. You have to attach the coil of wire to a battery to pass a current through it. To make the electromagnet really strong you have to leave the metal inside the coil of wire. <u>It doesn't matter what metal it is</u> [1 mark]. Adding more turns to the solenoid will make it stronger too. <u>Changing the size of the current won't make any difference to how strong the electromagnet is</u> [1 mark]. Once you have made the electromagnet work <u>it will keep on working even when you've switched the current off</u> [1 mark].

4. a) Place the plotting compass near the wire. [1 mark]

Check the direction of the compass needle with no current. [1 mark]

Switch on the current and see if the compass needle changes direction. [1 mark]

b) Place the plotting compass on the card. Move it around the wire and mark on the card the alignment of the compass needle at each position. [1 mark]

Draw curved lines to link the marks. [1 mark]

Add arrows on the field lines in the direction the compass needle points. [1 mark]

5.

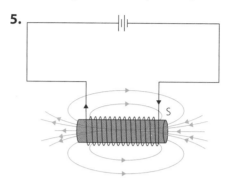

a) Evenly spaced straight field lines inside the solenoid [1 mark]; correct shape of field outside the solenoid [1 mark].

b) Correct direction shown for field lines inside the solenoid [1 mark]. Correct direction shown for field lines outside the solenoid [1 mark].